I0488167

JUNE 2014

electronics

SOUTH ASIA'S MOST POPULAR ELECTRONICS MAGAZINE

FOR YOU PLUS +

TOP 5
DO-IT-YOURSELF

- Get Set Go with Raspberry Pi Camera
- Brushless DC Motor Driver
- Alcohol Level Tester
- Power Factor Corrector
- Designing with FPGAs: FPGA-Embedded Processors

Plus, many more make your own projects inside

3D Printers

Technology That Changes Everything

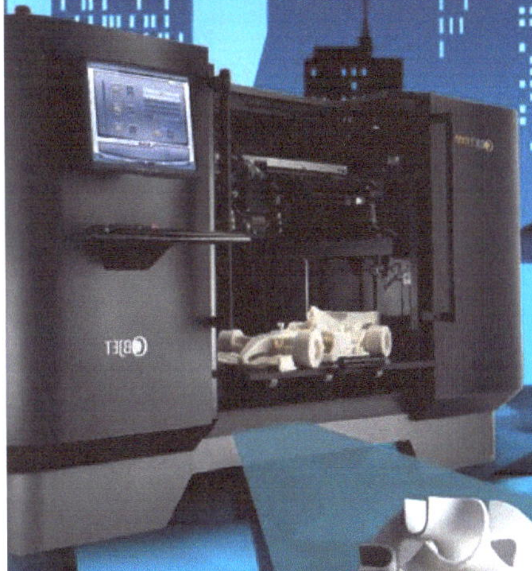

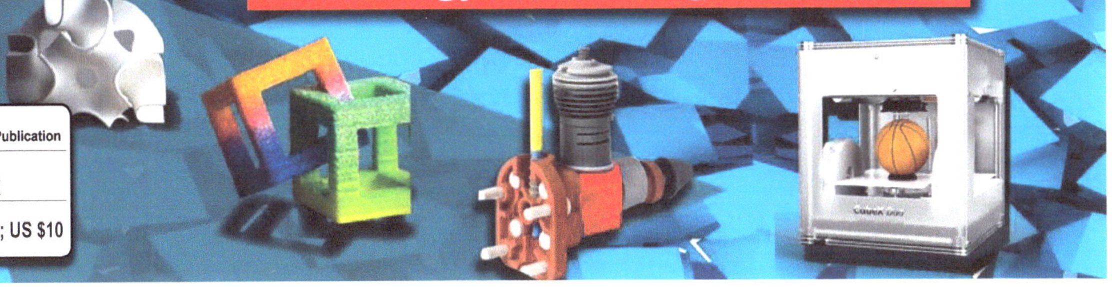

An **EFY**GROUP Publication
Vol. 3 No. 2
ISSN 0013-516X
Pages: 150+8 | UK #5; US $10

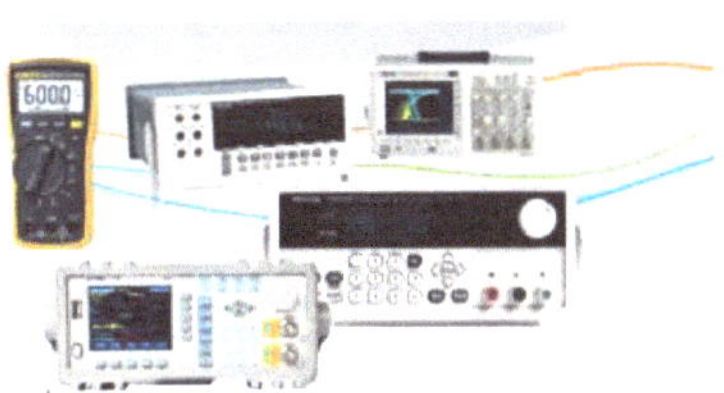

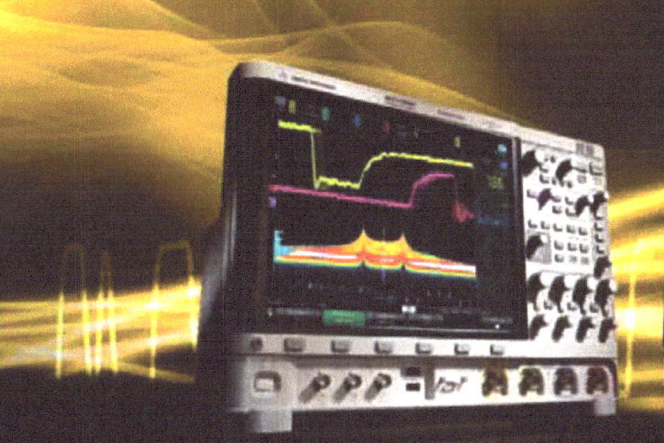 

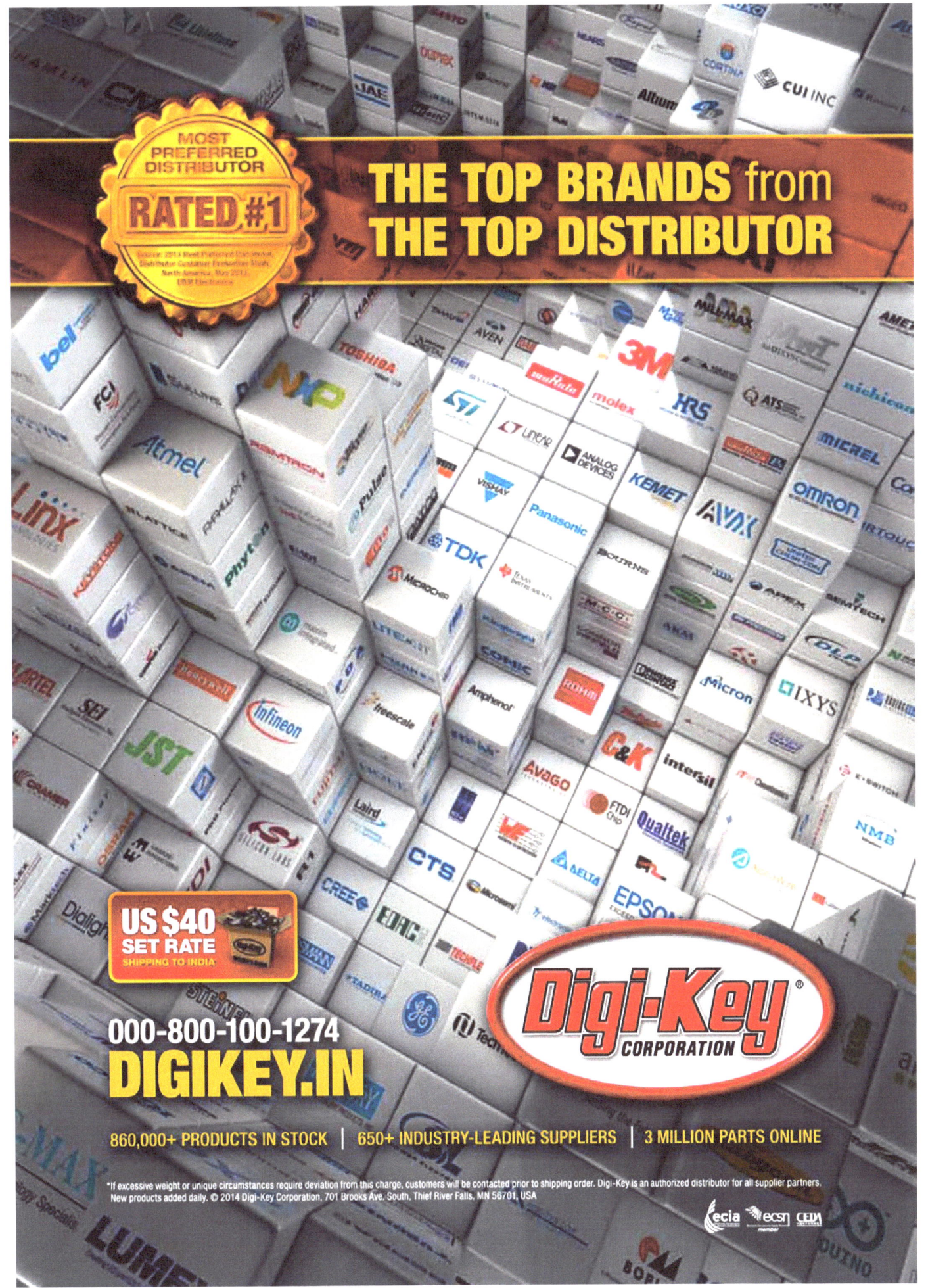

contents

ELECTRONICS FOR YOU PLUS | JUNE 2014 | VOL. 3 NO. 2

Hands-On

EDITOR : Ramesh Chopra

EDITORIAL CORRESPONDENCE : Editorial Secretary Phone: 011-26810601; E-mail: editsec@efyindia.com (Technical queries: efylab@efyindia.com)

SUBSCRIPTIONS & MISSING ISSUES : Phone: 011-26810601 or 02 or 03 E-mail: support@efyindia.com

BACK ISSUES, BOOKS, CDs, PCBs etc. : Kits'n'Spares, New Delhi Phone: 011-26371661, 26371662 E-mail: info@kitsnspares.com

EXCLUSIVE NEWSSTAND DISTRIBUTOR : IBH Books & Magazine Distributors Ltd, Mumbai Phone: 022-40497401, 40497402, 40497474, 40497413; Fax: 40497434 E-mail: circulations@ibhworld.com

ADVERTISEMENTS NEW DELHI (HEAD OFFICE) : Ph: 011-26810601 or 02 or 03 E-mail: efyenq@efyindia.com

MUMBAI : Ph: 022-24950047, 24928520 E-mail: efymum@efyindia.com

BENGALURU : Ph: 080-25260394, 25260023 E-mail: efyblr@efyindia.com

CHENNAI : Ph: 09916390422 E-mail: efyenq@efyindia.com

HYDERABAD : Ph: 09916390422 E-mail: efyenq@efyindia.com

KOLKATA : Ph: 08800094202 E-mail: efyenq@efyindia.com

PUNE : Ph: 09223232006 E-mail: efypune@efyindia.com

GUJARAT : Ph: 09821267855 E-mail: efyahd@efyindia.com

CHINA : Power Pioneer Group Inc. Ph: (86 755) 83729797, (86) 13923802595 E-mail: powerpioneer@efyindia.com

JAPAN : Tandem Inc., Ph: 81-3-3541-4166 E-mail: tandem@efyindia.com

SINGAPORE : Publicitas Singapore Pte Ltd Ph: +65-6836 2272 E-mail: publicitas@efyindia.com

TAIWAN : J.K. Media, Ph: 886-2-87726780 ext. 10 E-mail: jkmedia@efyindia.com

UNITED STATES : E & Tech Media Ph: +1 860 536 6677 E-mail: veroniquelamarque@gmail.com

NEXT ISSUE • Raspberry Pi   • PCB Industry in India   • Budget Friendly Oscilloscopes

Unrivaled Performance, Flexibility,
and Value for Automated Test

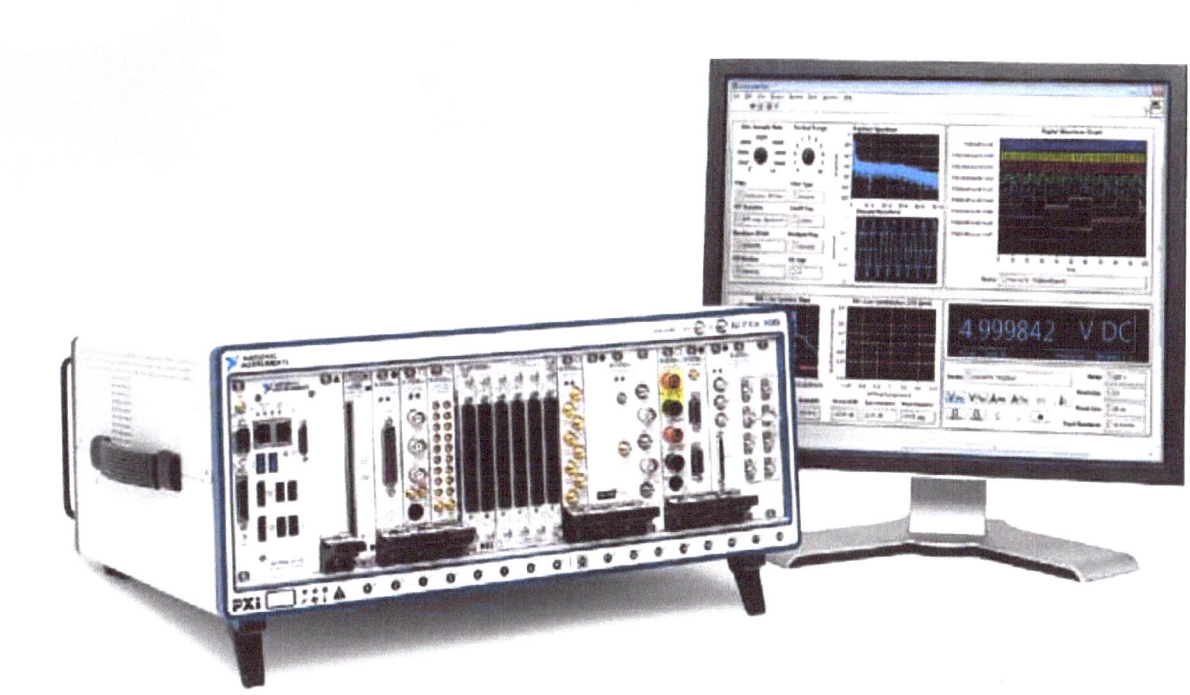

NI PXI hardware, combined with NI LabVIEW system design software, delivers increased performance, flexibility, and value. Using this combination of modular hardware and productive software, engineers have dramatically reduced costs, realized faster test execution, improved throughput, and increased scalability. With more than 500 PXI products, over 200 global locations, and 700+ Alliance Partners, NI offers the only complete solution for your ever-changing automated test needs.

LabVIEW helps you program the way you think—graphically—while simplifying your approach with built-in analysis and unmatched hardware integration.

>> Accelerate your productivity at ni.com/automated-test-platform

Micromax Launches Smartphone Supporting 21 Languages

Powered by quad-core processor and running on Android OS

Micromax recently launched its Micromax Unite 2, the world's first smartphone to support 21 different languages.

The device is powered by a 1.3GHz quad-core processor and runs on Android KitKat 4.4.2 operating system. It will be available in grey, white, red and green colours.

The smartphone has an 11.9cm (4.7-inch) Bright Graph IPS display with a resolution of 800x480. It runs on a 2000mAh battery and has a 5-mega-pixel rear camera with flash. In addition, it has a 2-mega-pixel front camera as well.

Price: ₹ 6999

BenQ Launches Full-HD Gaming Monitors

Featuring exclusive Revolution Eyes technology for exceptional performance

BenQ has launched XL-Z series of first-person shooter (FPS) purpose-built gaming monitors: the 68.6cm (27-inch) XL2720Z and 61cm (24-inch) XL2420Z and XL2411Z widescreen full-HD displays.

Featuring BenQ's exclusive Revolution Eyes technology for exceptional monitor performance, the displays feature motion blur reduction, Low Blue Light technology, zero flicker, gaming refresh rate optimisation management (GROM), and 1ms GTG response times. In addition, the new models offer BenQ's latest firmware, which enables third-party utility support for further optimisation options. The end result is striking visual clarity, seamless motion fluidity, and added viewing comfort for hours of action-packed game play.

The FPS mode featured in XL-Z series lets players tap into the fundamental insights of pro gamers and view the game the way a gaming legend would see it. This feature automatically adjusts monitor calibrations to provide users with optimal brightness, contrast, sharpness and colour tint.

Featuring Low Blue Light technology, the XL-Z monitors successfully manage the exposure of blue spectrum light emitted, resulting in more comfortable viewing. To help gamers protect their eyes during extended periods, the monitors provide several adjustable low blue light levels that automatically adjust emission without affecting quality.

With the GROM management system, gamers also gain the freedom to custom-build their personal gaming experience according to viewing preferences such as refresh rates (100/120/144Hz), display resolutions and screen sizes.

Price: XL2720Z: ₹ 37,500
XL2420Z: ₹ 32,500
XL2411Z: ₹ 30,000

Samsung's New Curved TVs with 4k Resolution

The world's first curved UHD television

Samsung Electronics has launched its curved TV range in India. Amongst the devices showcased recently was the world's first curved UHD television.

With the launch of world's first curved UHD TV, Samsung is blending its innovative curved form factor with its UHD TV technology. Samsung's new curved televisions have four times the resolution and pixels of full HD, which means 4k resolution.

In addition, the TVs support HEVC, HDMI 2.0, MHL 3.0 and HDCP 2.2. The devices launched include the Samsung 105 Curved UHD TV, HU9000 Curved UHD TV, H8000 Curved Smart LED TV, HU8500 UHD TV and the HU7000 television.

Price: ₹ 100,000 to ₹ 450,000 depending on screen size

Performance you need at a price you can afford

EXCLUSIVE OFFER

Handles everyday test challenges without challenging your budget

NEW

TBS1000B/EDU Series - valid upto 27th Oct 2014

**VAT/Taxes Apply

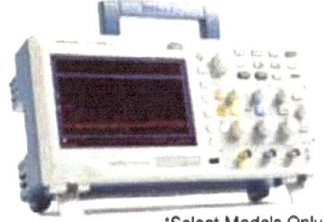

*Select Models Only

TBS1000B Series
Designed for extensive monitoring and analysis activities

- ✓ 7-inch WVGA display
- ✓ Unique TrendPlot™ feature
- ✓ Enhanced limit-test feature
- ✓ Digital real-time sampling
- ✓ 5-year warranty

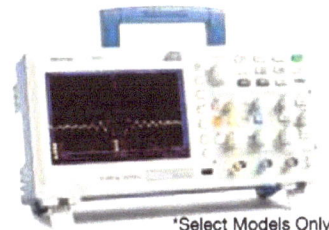

*Select Models Only

TBS1000B-EDU Series
The world's first dedicated Teaching Oscilloscope

- ✓ 7-inch WVGA display
- ✓ PC based courseware editor
- ✓ Web-based content eco-system
- ✓ Digital real-time sampling
- ✓ 5-year warranty

Model	Bandwidth	Sample Rate	Record Length	Channels	Price
TBS1072B	70MHz	1GS/s	2.5k point	2	Rs. 29000
TBS1072B-EDU	70MHz	1GS/s	2.5k point	2	Rs. 29000
TBS1102B	100MHz	2GS/s	2.5k point	2	Rs. 39000
TBS1102B-EDU	100MHz	2GS/s	2.5k point	2	Rs. 39000
TBS1152B	150MHz	2GS/s	2.5k point	2	Rs. 49000
TBS1152B-EDU	150MHz	2GS/s	2.5k point	2	Rs. 49000
TBS1202B	200MHz	2GS/s	2.5k point	2	Rs. 59000
TBS1202B-EDU	200MHz	2GS/s	2.5k point	2	Rs. 59000

For more information, please call our sales channel partner in the city nearest to you. **Ahmedabad :** Optimized Solutions Pvt. Ltd. (91-79) 30080808 / 26589008; **Bangalore:** Aarjay International (91-80) 43409200, Convergent Technologies (91-80) 23490111 / 179; **Chennai :** Aarjay (91-44) 40509200, Primetech Instruments Pvt. Ltd. (91-44) 24492961; **Andhra Pradesh / Orissa :** Peridot Technologies (91-40) 23405855; **Indore :** Optimized Solutions Pvt. Ltd. (91) 9909979950; **Kerala :** Convergent Technologies (91-471) 2462412 / 386; **Kolkata :** Techno Scientific : (91-33) 32948567; **Mumbai :** Vitronics India (91-22) 28506037; **New Delhi :** S.P.I. Engineers Pvt. Ltd. (91) 9810157421; Convergent Technologies (91-11) 42481121 / 1131; **Pune :** Cyronics Instruments Pvt. Ltd. (91-20) 24208200; **North East India :** Vishal Vyapar Vikash (91-361) 2540701.

Enquires Email To : india.mktg@tek.com

Tektronix

GizMo ByTes

Mobile App for Real-Time Electric Meters' Readings

Mobicule Technologies together with Riverside Utilities has developed and launched its new mobile application msales Utility app. The app will ensure faster, smarter and authentic electric meter reading and billing system for the consumers. All the meter readers need is a smartphone with GPRS connection. They can note down the real-time reading, click the image of the same with geo-tagging feature and submit it to the server directly by clicking on the submit button.

USB Portable Charger from Sony

Sony India has launched its new CP-V3A USB portable charger with a capacity of 3000mAh. The device is available in six different colours and is priced at ₹ 1590. It contains Sony's lithium-ion battery and uses the company's own Hybrid-Gel technology. The technology allows the charger to retain 90 per cent of its capacity till 1000 charges. It gives a high output of 1.5A for a speedy charge.

Intex Launches New Smartphone

Intex has introduced its new Aqua i5 HD smartphone in the market. It sports a 12.7cm (5-inch) 720-pixel HD display and is powered by a 1.3GHz quad-core processor. It runs on Android 4.2 Jelly Bean operating system, which can be updated to Android 4.4.2 KitKat in future. The device also has a 13-mega-pixel camera on the back with LED flash along with a 5-mega-pixel front camera. Its 1GB of RAW and 4GB of internal storage can be extended using a micro-SD card. It runs on a 2000mAh battery. The device has been priced at ₹ 9990 and will be available in black and white colours.

Cubix Music Powerhouse that Fits in Your Palm

Produces very powerful yet pure sound

Portronics is proud to present Cubix, a compact and powerful speaker that is stylish and ultra portable. It is an extremely bag-friendly portable speaker that will make up for that distinct lack of audio quality that your devices' speakers cannot provide.

The sound output of Cubix is very powerful and pure. It not only amplifies the sound but also makes it more melodious and audible. So whether you are in the comfort of your home, a garden or in a hotel room, Cubix will give you pristine and clear sound.

Cubix offers great adaptability and can easily be played with your

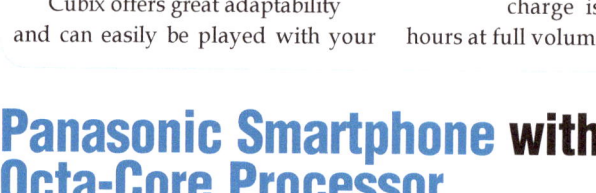

Price: ₹ 1299

smartphones, tablets, MP3 players, notebooks, desktops, hand-held gaming consoles, or any other audio-enabled device. The most unique feature of Cubix is that it needs no separate power source to run. Just connect the speaker to an Aux port of the device and you are ready to enjoy.

Cubix comes in three silicon casing colours of blue, orange and yellow. Being small, it can easily be accommodated in the palm of your hand, or fit comfortably in your pocket, purse or your laptop bag. The play time of the speaker in a single charge is more than five hours at full volume.

Panasonic Smartphone with Octa-Core Processor

Has 13MP rear camera with auto-focus and LED flash

Panasonic has announced the launch of P81 smartphone with octa-core processor.

The Panasonic P81 packs a 14cm (5.5-inch) 720p HD display. The device offers dual-SIM support and is powered by a 1.7GHz octa-core MediaTek processor. It packs 1GB of RAM and runs on Android 4.2.2. The P81 has a 2500mAh battery, 8GB of on-board storage, microSD card slot and the usual connectivity options.

On the camera front,

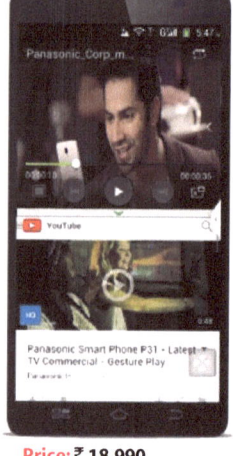

Price: ₹ 18,990

it has a 13MP rear camera with auto-focus and LED flash, along with a 2MP front shooter. Panasonic has added quite a few software tweaks in the device, including the device cleaner, dual window support, Pop-i Player and gesture support.

Panasonic P81 is packed with freebies worth ₹ 11,900, including six-month subscription to Evernote premium, free downloads from Hungama for a month, and accessories such as flip cover and screen protector.

 **HONGFA RELAY**

 **MILLENNIUM SEMICONDUCTORS**

THE LEADING RELAY R&D AND MANUFACTURING BASE GLOBALLY

Xiamen Hongfa Electroacoustic Co. Ltd is the first largest relay manufacturer in China and one of the leading relay manufacturers and suppliers in the world. It produces a wide range of products, including relays, low-voltage devices, precision components and automatic equipments. In particular, it has over 160 types of relay categories, including power relays, latching relays, automotive relays, signal relays, industrial & electrical appliance relays, safety & electrical appliance relays, solar relays and hermeticallysealed relays, which provide over 40,000 regularly-used relay specifications.

 **POWER RELAY**

UPS, Inverter, Control panel, Voltage stabiliser

HF3FA (215)
- 1 pole 15A 125VAC, 10A 250VAC/28VDC
- Product in accordance to IEC 60335-1 available

HF3FF (215)
- 1 pole 10A 277VAC/28VDC

HF14FW(136)
- TV-8, 20A switching capability (1 pole configuration)
- 4kV dielectric strength
- Meeting VDE 0700, 0631 reinforce insulation

UPS, Inverter   Railways and Air-con.   Timers/Relay module

HF105F-1(136)
- 40A switching capability
- 4kV dielectric strength (coil to contacts)
- Unenclosed, Plastic sealed and dust protected types available

HF102F
- Heavy load up to 5000VA
- Ideal for motor load (80A High inrush current)
- PCB & QC layouts available

HF33F
- 1 pole 10A 125VAC, 5A 250VAC/30VDC
- Creepage distance 8mm (NO to NC)
- High sensitivity 200mW available

Lighting application   UPS   PLC Controller   Air-con.

HF46F
- 1 pole 5A 277VAC/30VDC at 85℃
- Surge withstand voltage up to 10kV
- High sensitivity 200mW

HF115FD
- 16A switching capability
- Creepage distance 10mm
- Updated version of HF115F 1 pole equipped with automatic production line

HF49FD
- 3kV dielectric strength between coil and contacts
- Slim size width 5mm, height 12.5mm
- High sensitive min 120mW

HF118F(136)
- 1 pole 10A 2田 (2 pole) 5A
- dielectric strength 5kV
- Creepage distance >8mm
- Reflow soldering available

Telecom, EPABX, Mobile tower

HFD27
- High contact capacity 2A 30VDC
- Matching 16 pin IC socket
- Epoxy sealed for automatic-wave soldering and cleaning

HFD23
- 1 Form C configuration
- 2A switching capability
- High sensitive 150mW
- Plastic sealed type available

HFD3
- Surge withstand voltage up to 2500VAC, meets FCC Part 68 and Telecordia
- SMT and DIP types available
- Single side stable and latching type available

Control panel

HF18FF / HF18FH
- 7A switching capability (2C, 3C type)
- Various terminals, test button available
- Gold plated contact available
- 2 to 4 pole configurations

HF13F
- 15A switching capability (1 Form C)
- Conform to the CE low voltage directive
- 1 & 2 pole configurations
- Various terminals available

Gaming Hardware: Put Your Game Face On

The idea of gaming stopped being something for children a long time ago. PC and console games today are highly advanced in all respects, be it the storyline, content, graphics or the kind of hardware they use. That said, playing the newest games requires one to be equipped with top-of-the-line hardware, starting from gaming consoles to keyboards and mice. It is a huge market not only in India but all over the world, and here's what it has to offer

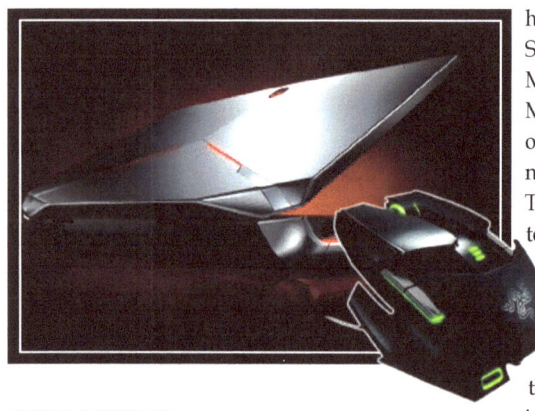

PRASID BANERJEE

Gaming hardware covers a wide array of items that gamers like to use to make their experience exhilarating. While the most common are gaming consoles, graphics cards and headsets, there is a lot more to it. Contrary to what many think, hardcore gamers still consider the PC to be the king of gaming. While consoles like Sony Playstation and Microsoft Xbox have for long gained immense popularity in the market, it's hard to overlook the raw power of top-of-the-line graphics cards like those sold by AMD and other giants in this segment.

Let us look at what all a gamer buys. Even though it isn't the undisputed winner, a gaming console is still an important piece of gaming equipment. Then there are graphics cards, headphones, mice and other things.

Gaming consoles

With open source giant Valve's Steam Machines close to making an appearance, the console arena is now more interesting than ever. While the battle has for long been between Sony Playstation (PS) and Microsoft Xbox, the Steam Machine threatens both of these in a way that has never been done before. The other option is the Nintendo Wii, which is more of a motion-driven gaming option.

The rule of thumb in this segment is to buy the newest version available in the market. But there is a small window after the launch of a new console, when you can still buy the older model. For example, both Microsoft and Sony launched their Xbox One and Playstation 4 consoles around October last year. So the window to buy the PS3 and Xbox 360 is just about over now. It takes some time for a new console to pick up speed and become the platform of choice for game makers, which is when you can buy the older version. But if you intend to buy a PS2 now, it would be a mistake, to say the least.

Also available in this segment are handheld gaming consoles. These, in our opinion, make more sense for less serious gamers. If you do not go past the 'easy' difficulty level on a game, but like to play it till the very end, you should get one of these. You can play your games on the go and it doubles up as a tool for listening to music and as a portable storage device as well.

Laptops

When we say that PC is still the king of gaming, it includes the important position held by gaming laptops. A lot of gamers like to buy gaming laptops instead of assembling an entire PC unit. The portability factor plays a huge role in this. The leading name in this segment would be Alienware from Dell, but there are also other offerings from Asus etc.

Gaming laptops though are quite costly, with price tags of ₹ 100,000 and above. The units sold by Asus are somewhat cheaper as compared to Alienware, but Alienware has for long been the name in the gaming segment. If you are buying a gaming laptop, get ready to spend a bomb, because these are top-of-the-line devices that exude raw power to give you the best gaming experience.

On the other hand, a recreational yet serious gamer (like those who spent money on every new game, but spend just enough time to finish the game in the easy mode) could also go for a top-of- the-line laptop like the Dell XPS series. These aren't meant for gaming, but they are powerful enough to play many of the best games.

Graphics cards

Here we enter the realm of hardcore gamers. A high-end graphics card that can support almost any game isn't available for chump change. The nVidia GeForce GTX series, for example, can cost ₹ 20,000 and more. To put it in perspective, you can assemble a low-end PC for a little more than that.

A serious and true gamer will not compromise on this bit of hardware and will spend lots on it. So, whether you choose a graphics card from nVidia, AMD or anyone else, get ready to spend some big bucks on this. What you are looking for is something that suits the kind of games you like to play

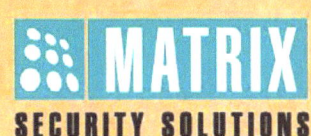

GAMING HARDWARE TO CHOOSE FROM

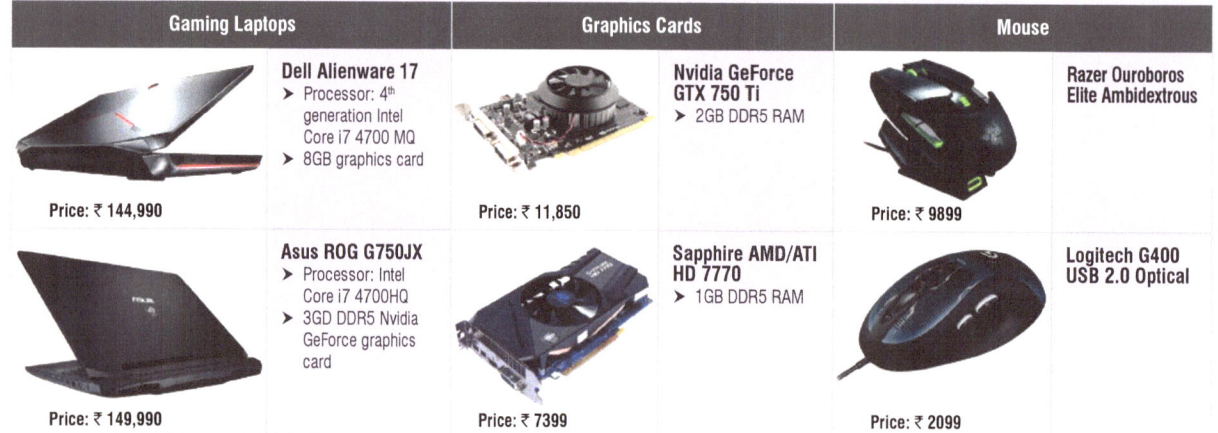

Gaming Laptops	Graphics Cards	Mouse
Dell Alienware 17 ➤ Processor: 4th generation Intel Core i7 4700 MQ ➤ 8GB graphics card Price: ₹ 144,990	**Nvidia GeForce GTX 750 Ti** ➤ 2GB DDR5 RAM Price: ₹ 11,850	**Razer Ouroboros Elite Ambidextrous** Price: ₹ 9899
Asus ROG G750JX ➤ Processor: Intel Core i7 4700HQ ➤ 3GD DDR5 Nvidia GeForce graphics card Price: ₹ 149,990	**Sapphire AMD/ATI HD 7770** ➤ 1GB DDR5 RAM Price: ₹ 7399	**Logitech G400 USB 2.0 Optical** Price: ₹ 2099

and is still within your budget. The nVidia GeForce GTX 750 Ti is a good option that costs within ₹ 15,000.

Keyboards

Also part of a gamer's shopping list are gaming accessories, like mouse, keyboard and headphones. There are some brands, like Razer and Steelseries, which are particularly associated with gaming. Compared to graphics card, these are much cheaper but, compared to your ordinary headphones, a gamer's gear still costs good money.

A gaming keyboard can cost anywhere between ₹ 1000 to ₹ 10,000. You would make a good investment with something like a Cooler Master Devas-tator or a Logitech Gaming Keyboard G105, which cost between ₹ 3000 and ₹ 4000. On the other hand, if you have the budget, you can go for a Razer Blackwindow Ultimate 2013 keyboard, which costs nearly ₹ 10,000.

For a part-time gamer, the keyboard costing around ₹ 1000 to ₹ 2000 would do the trick. The only advice

GAMING HARDWARE TO CHOOSE FROM

Consoles	Keyboards
Sony Playstation 4	**Razer Deathstalker Ultimate USB**
Price: ₹ 39,000	Price: ₹ 23,000
Microsoft Xbox 360 Kinect Holiday Bundle	**Steelseries Shift USB 2.0**
Price: ₹ 26,490	Price: ₹ 3999

we can give you is, buy online to enjoy good discounts.

Mouse

Many non-gamers do not understand the logic behind spending ₹ 3000 on a gaming mouse. Well, for a gamer, a good mouse, with easy-to-adjust DPI (dots per inch) settings, additional keys, etc is almost an imperative. The usual features of a gaming mouse include adjustable DPI settings, programmable buttons, acceleration, onboard memory and, in some cases, adjustable weight.

Out of the above-mentioned reasons, the first two are perhaps the most useful. DPI decides how much a mouse moves in relation to the movement of your hand. So, more the DPI, more it moves with less movement of the hand. So, if you play a lot of first-person shooters and are usually a sniper, then low DPI settings is what you would go for. High DPI would be useful if you have a large monitor and/or are playing fast-paced run-and-shoot games. Example: the Cooler Master Spawn, which costs ₹ 3000 and comes with 3500 DPI and seven programmable buttons.

Programmable buttons allow you to store key combinations. You can also use them as alternatives for keyboard keys. These are usually found in mid- and high-end gaming mice. A part-time gamer may often

compromise on this aspect in order to save money.

Headphones

Noise cancellation is the order of the day here. Most gaming headsets come with the noise cancellation feature, making sure that when you are playing, it is just you and your game; the entire world can take a number and stand in line.

Buying gaming headsets is really a matter of choice. You may want to spend ₹ 1000 or ₹ 5000, or even more on a headset. Alternatively, you may want to spend some more and get a good sound system, which helps you for other activities too. The choice is yours. Many gamers buy headsets for outdoor gaming, like when they are out in public or playing at a fellow gamer's place.

Also, headsets are really useful when you are chatting with other players on multiplayer games and taking or giving commands.

Conclusion

Gamers can really be divided into two categories—serious and part timers. Both make for big markets for the producers. If you are going to buy gaming hardware, the only worthwhile advice is to buy the best you can in the budget that you have. The task is much easier for a part-time gamer than it is for the serious ones. ●

The author is a correspondent at EFY

Your clothes may soon change colours on touch

How would you like to wear fabric that could change colours based on sound or touch? Well, the technology now exists. A research called Chromosonic, conducted by Judit Eszter Karpati from Hungary, has made this possible. According to reports, Karpati's efforts to explore how digital media can be mixed with textile medium has resulted in the building of such a medium, which can be called Electronic Programmable Textile Interface.

Karpati apparently used an Arduino with 12V power supply and 20 custom PCB chips, which control four industrial 24V DC power supplies. These tools heat up the two handmade textiles, which have been embedded using nichrome wires and screen-printed thermo-chromatic dying.

According to reports, the fabric changes its colour based on sound and touch. Many have already said that wearing this could make the wearer look like a chameleon, although the technology would perhaps need a lot more work before that kind of perfection is achieved.

Please scan the QR code (see pic) that describes the project and will help you understand more about it.

Your body heat to charge your mobile phone

In a bid to ensure a stable and reliable power supply to your ever-power-hungry smartphone, a team of researchers has now come up with the idea of a glass-fabric-based thermoelectric (TE) generator. The TE generator will be an extremely light and flexible solution to powering up heart monitors, smart glasses and other wearable tech, producing electricity from the heat emitted from a human body.

"Mobile phones consume high electrical energy compared to electri-

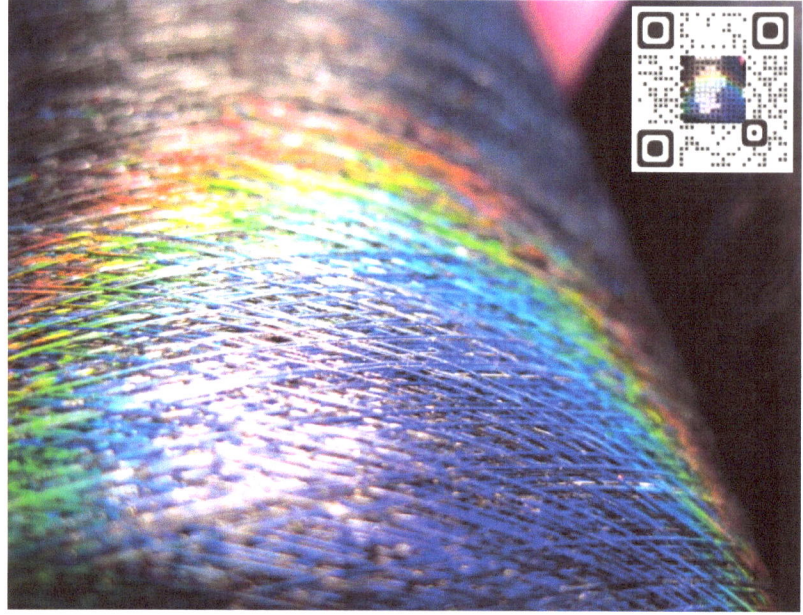

Temperature-sensitive fibre changing colour

cal sensors. Right now we are trying to make a sample that provides electricity for medical sensors," Byung Jin Cho, a professor of electrical engineering at Korea Advanced Institute of Science and Technology in South Korea was quoted as saying. "After that, smartphones will be the next application of the TE generator," he added.

The flexible TE generator could minimise thermal energy loss while maximising power output. It has a self-sustaining structure, eliminating thick external substrates (usually made of ceramic or alumina) that hold inorganic TE materials, and efficiently uses the small yet significant temperature difference between skin and air to create power.

The researchers have developed two types of TE generators. The organic-based TE generators make use of polymers that are highly flexible and compatible with human skin. Although ideal for wearable electronics, these have a low power output. Inorganic-based TE generators, however,

produce a high electrical energy. The downside is they are heavy, rigid and bulky. Cho's technology successfully combines the best of both organic and inorganic tech.

Method to commercialise graphene for electronics revealed

Samsung has discovered a synthesis method to speed up commercialisation of graphene, a unique material ideally suited for electronic devices. Samsung Advanced Institute of Technology (SAIT) has joined hands with Sungkyunkwan University to be the first in the world to develop this new method.

Graphene has one hundred times greater electron mobility than silicon, which is the most widely used material in semiconductors today. It is more durable than steel and has high heat conductibility as well as flexibility, which makes it the perfect material for use in flexible displays, wearables and other next-generation electronic devices.

SAIT revealed a new method of

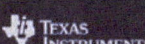  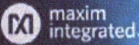 
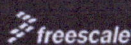  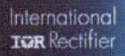    
 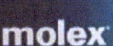  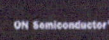 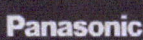
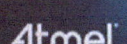 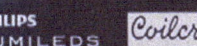 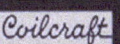 

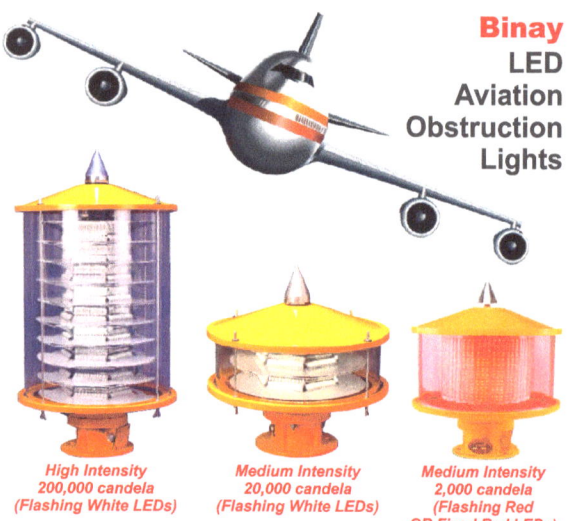
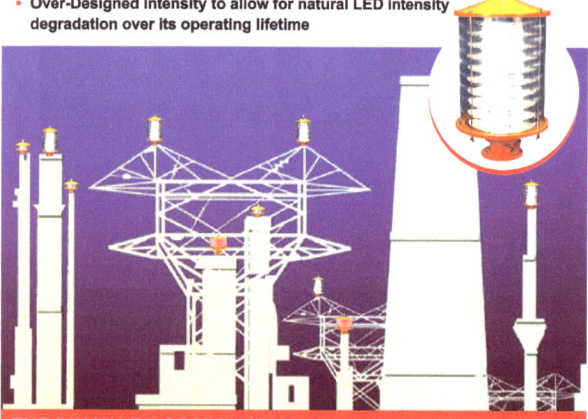

growing large-area, single-crystal, wafer-scale graphene. Engineers around the world have invested heavily in research for the commercialisation of graphene, but have faced many obstacles due to the challenges associated with it. The new method synthesises large-area graphene into a single crystal on a semiconductor, maintaining its electric and mechanical properties.

New lithium battery created in Japan

The long life of lithium-ion batteries makes them the rechargeable choice for everything from implantable medical devices to wearable consumer electronics. But lithium-ion batteries rely on liquid chemistries involving lithium salts dissolved in organic solvents, creating flame risks that would be avoided if the cells were completely solid state.

Now a team of researchers at Tohoku University in Japan has created a new type of lithium-ion conductor for future batteries that could be the basis for a whole new generation of solid-state batteries. It uses rock salt lithium borohydride ($LiBH_4$), a well-known agent in organic chemistry laboratories that has been considered for batteries before, but up to now it has only worked at high temperatures or pressures.

In the journal APL Materials, from AIP Publishing, the researchers describe how they doped a cubic lattice of KI molecules with the $LiBH_4$. This allowed them to stabilise the high-pressure form of lithium borohydride and make a solid solution at normal atmospheric pressure that was stable at room temperature.

Gesture recognition and virtual touch for vehicle controls

A new automotive cockpit concept demonstrates that futuristic-advancements, such as 3D gesture recognition, are now closer to reality. The Horizon cockpit concept, which Visteon has been demonstrating to global vehicle manufacturers, blends three emerging technologies to transform the way a driver controls features such as interior temperature, audio and navigation.

Gesture control, an advanced camera system, maps the user's hand to replicate a virtual hand on the centre stack. Horizon drivers can then easily operate certain vehicle controls simply by moving their hand or just a finger, without making contact with the instrument cluster. For example, radio volume can be adjusted by making a turning motion with one's hand.

By integrating a pressure-sensitive touch pad with a virtual touch screen, drivers can operate centre-stack controls without having to physically reach for them. The touch pad recognises the amount of pressure applied for improved responses. The touch pad can be implemented with any soft material, such as leather or cloth, allowing flexibility for its location.

High-resolution graphics present information on two separate planes, bringing only those controls with which the driver is interacting to the forefront. The driver can virtually 'push' through the graphics on the first pane to access the second.

Coming soon: robots that can touch and feel

Indian-origin researcher Dr Ravinder Dahiya, a professor of electronic and nanoscale engineering, and Professor Duncan Gregory, chair in Inorganic Materials in the School of Chemistry at the University of Glasgow have now ventured on quite an adventure: to make robots go all touchy-feely, that is! They are currently working on a technology to create an ultra-flexible tactile skin for robots that could bring them a step closer to being more human!

The researchers are looking to combine nanotechnology with robotics and develop a printing technique for high-mobility materials such as silicon. Once successful, they will then embed electronics and sensors on a bendable silicon-based surface approximately 50 micrometres thick, or even less. The researchers plan to cover the robot's entire body, so it can feel and touch throughout rather than just restricting to its hands.

Robots that can touch and feel

Likewise, Engineered Arts Limited has come up with a new robot that features a dynamic face reflecting varied moods and even responding accurately to expressions of individuals giving it a look. SociBot-Mini is based on a fully-integrated LED pico-projection technology featuring custom-developed optics.

SociBot-Mini can find and track faces, and also accurately tell an individual's gender and estimate his/her age. These are achieved through an-image processing software called SHORE. The robot makes eye contact without having to be prompted and attends to users' needs by analysing their body language. The SociBot features an HD RGB camera, a fully-articulated neck, an IR depth sensor, and sports high-quality audio and a public-facing touchscreen interface.

Nickel-fluoride film to power flexible electronics

Researchers have taken the first step towards creating bendable and fully-foldable electronic devices. In a bid to replace the traditional polymers and carbon-based materials, such as carbon nanotubes, deployed currently to provide critical back-up power for portable and flexible electronics, researchers at the University of Houston have now developed a new 'thin' film for energy storage. The film is much thinner than paper, is flexible as well as bendable and can store enough energy to power bendable and fully foldable electronic devices in the near future.

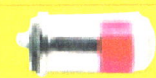
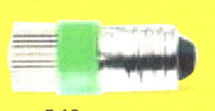

James Tour and his colleagues have developed gold nanomesh electrodes that offer ultra-high stretchability. At the same time, they can provide good electrical conductivity. According to Tour and Co, the current crop of materials fall short as reliable supercapacitors, and their new technology is well on its way to become the viable alternative. "While flexible gadgets such as 'electronic skin' and roll-up touchscreens are moving ever closer to reality, their would-be power sources are either too wimpy or too stiff. But that's changing fast," the researchers were quoted in the Journal of the American Chemical Society (ACS).

Tour's 'flexible' material consists of nanoporous nickel-fluoride electrodes layered around a solid electrolyte providing battery-like super-capacitor performance. These nano-pores allow ions to flow easily throughout the material, so much so that the resulting structure can pack in far more power for its size. The material is thus able to provide the best qualities of a high-efficiency battery as well as a high-powered super-capacitor. Researchers claim that you can bend and fold the film and recharge it thousands of times with little or no loss in performance.

Have some lettuce... grown in Fujitsu's Fab!

Are you a food connoisseur? Even if you are not, how would you like to taste a patch of iceberg lettuce grown in a hi-tech clean room powered by Internet of Things smartness?

Semiconductor-firm Fujitsu began selling its first round of 'Kirei Yasai' lettuce on May 7th this year. The salad ingredient is grown in a 2000 square metre facility that previously used to fabricate microchips and other components for electronics products.

These lettuce are the result of a lot of sensor-based arrays and systems working in tandem with cloud computing solutions to crunch data and ensure that the environment remains at the optimal levels throughout the growth process. This includes monitoring temperature, humidity and fertiliser composition as they try to figure out the best growing conditions and ways to control micro organisms.

The project uses 'Akisai,' the food and agricultural cloud from Fujitsu that leverages ICT to dramatically improve the efficiency of agricultural operations. Akisai will be used to continually analyse agricultural data for highly productive cultivation and to facilitate the entire management process, including management, production and sales, for a more efficient agricultural operation.

Are birds getting lost because of our electronics?

This debate is over. And the answer is 'Yes.' A letter published online in the journal 'Nature' goes to prove that migratory birds are unable to use their magnetic compass in the presence of manmade electromagnetic noise around them.

The test was conducted on European Robins, by exposing them to electromagnetic noise present in unscreened wooden huts at the University of Oldenburg campus. Upon being exposed, these birds were unable to use their internal magnetic compass and lost orientation. When aluminium-screened huts were used, which attenuated the electromagnetic noise, their capability to orient themselves to the magnetic field reappeared.

The question now is: Does it affect biological processes and human health as strongly as it affects these poor birds?

Futuristic mind-controlled prosthetic arm is here

DARPA has finally succeeded in getting FDA approval for their prosthetic arm with near-natural control. Named the DEKA arm system, it is an advanced electromechanical prosthetic arm that allows almost full control of the arm via electrical signals from electromyogram electrodes.

It allows for simultaneous control of multiple joints using a variety of input devices, including wireless signals generated by innovative sensors on the user's feet. It can handle anything from sensitive items like grapes and eggs, all the way to power tools like a hand drill.

M2M technology to help turtles and tourism

Telit Wireless Solutions, a global provider of high-quality machine-to-machine (M2M) solutions, products and services, has announced that Turtle Sense, a project designed to accurately predict when sea turtles hatch and emerge from their nests, can integrate Telit HE910 modules to communicate sensor data. The goal

Telit HE910 modules to communicate sensor data

is to protect both the threatened and endangered sea turtles and the area's critical tourism industry. The Turtle Sense technology was developed by Nerds-Without-Borders.

Because coastal economies rely heavily on tourism from public access to beaches, business and environmental interests must balance the popularity of coastal towns with requirements for the preservation of sea turtle nesting areas prescribed in The Endangered Species Act. Currently, there is no reliable way to predict when the tiny turtles will emerge from their nests near the dune line and parade to the surf.

In order to extend battery life throughout the nesting, incubation and hatching period, transmission timing starts slow and intermittent and increases in frequency once activity is reported.

Check efytimes.com for more news, daily

SPC

YOU CAN DO BETTER WITH EVE BATTERIES

The EVE Patented Super Pulse Capacitor can store and discharge pulse current within a temperature range of - 40°C to + 85°C. it is the ideal power source for long-term standby and high-current pulse applications. EVE unique design is based upon the latest technology to ensure safety and reliability.

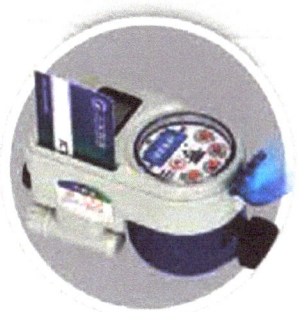

OVER
400 MILLION
UTILITY METERS HAVE BEEN MANUFACTURED WITH EVE BATTERIES, BETTER SOLUTIONS FOR BETTER METERS. LET EVE SUPPORT ALL YOUR POWER REQUIREMENTS.

Model	Dimension (metric) Diameter/Length	Capacity(As)	Maximum Discharge Current(mA)		Self Discharge Current(μA)		Available Terminations
			Continuous	Pulse	RT	80°C	
SPC 0920	9.0 X 21.0	30	150	500	2	5	S T 2PT 3PT
SPC 1520	15.1 X 21.0	140	500	2000	1	5	S T 2PT 3PT
SPC 1530	15.1 X 27.0	250	750	3000	1.8	8	S T 2PT 3PT
SPC 1550	15.1 X 51.0	560	2000	5000	3	15	S T 2PT 3PT

EVE ENERGY CO., LTD.

No.36,Huifeng 7th Road, Zhongkai Hi-Tech Zone, Huizhou, Guangdong, China
Tel : +86-752-2606966   Fax : +86-752-2606033
E-mail : sales@evebattery.com
Web : www.evebattery.com

Authorized Channel Partners in India
SM Electronic Technologies Pvt. Ltd.,
#1790, 5th Main, 9th Cross, RPC Layout, Vijayanagar
2nd Stage, Bangalore - 560040
Contact Person: Mr. Rajiv Anand, Mob : +91-9810715377
Tel : + 91-80 23301030   Fax : + 91-80 23387197
E-mail : sales@mysmindia.com

PERFECT SOLUTIONS INC.314, Krishna Mall,
Sector-12 Dwarka, New Delhi - 110075
Tel : +91-9213258888   Fax:+91- 45649530
web:www.psiindia.com
E-mail : sales@psiindia.com, sources2004@gmail.com

3D Printing: The Technology That Changes Everything

Ever since key 3D printing patents expired earlier this year, there has been media frenzy predicting a bigger printing boom about to happen. Let's take a look at the upcoming technologies for 3D printing, and how these could change our world

SNEHA AMBASTHA

The main benefit of 3D printing is the reduction in time, effort and energy taken to convert a design into a prototype. In a way, you can now build your products at home.

Older technologies like stereolithography (SLA or SL), selective laser sintering (SLS) and fused deposition modelling (FDM) or fused filament fabrication (FFF) are now available risk-free for use and sharing because their patents have expired. This has helped the open source community to increase their role in the enhancement of 3D printing technology.

Some time ago, the open source community took up a project called RepRap and popularised FDM 3D printers. Next in line, SLA and SLS are seen as the future of advanced technologies as their patents lapsed this February. This means that these technologies can now easily get into the hands of hackers or tinkerers, enabling them

to develop their own software and technologies, and then sell these printers at a lower price, popularising them to a much greater extent than at present.

Priya Kuber, managing director, Arduino India says, "I have used an open source 3D printer, Prusa Mendel, which is something on the lines of RepRap. What I found interesting was the software side of the 3D printing." Priya explains that Slicer is a highly mathematical tool to represent geometry of 3D printing. It slices the complete drawing into small layers and then converts them into a 3D printable format (.stl). "I also used the online tool called TinkerCAD that allowed me to feed an image and create a 3D image file out of it. Now you can feed that file into Slicer and then 3D print it," she says.

Form 1 and B9 Creator are two open source 3D printers where the latter is completely open source. Form 1 from Formlabs "uses stereolithography tech-

nology to achieve a professional print quality that plastic extrusion printers just can't match. A high-precision optical system directs a laser across a tank of liquid resin, solidifying layers as thin as 25 microns. The build platform pulls your model upwards, out of the tank," according to the company.

3D printing goes mobile

You can now get your designs fabricated without even being physically close to a printer. It is done through an 'app.' The app takes over the designing part, enabling a preview of the final product, and then gives the print command to the 3D printer available nearby. This is like enabling Google's Cloud Print functionality on your 3D printer.

Karan Chapekar, director, KCbots says, "Autodesk recently launched an app called 123d Catch for IOS7. It captures multiple pictures (about 40) that can be shared online or 3D printed." Engineers do not have to be near the 3D printer any more. They can prepare the code on their app even when they are travelling and give a print command to their 3D printer kept in their lab or at office. The only requirement is that the mobile phone running the app should be connected to the printer and the printer should be, in turn, connected to Wi-Fi.

What's happening to 3D printer vendors

There are major brands in 3D printing industry that have recently merged to give way to new technologies and help themselves grow in this field. Objet Ltd. and Makerbot Industries have merged together with Stratasys Inc., whereas Z Corp has merged with 3D Systems. 3D Nation has partnered with Sunglass and, very recently, Pirate3D tied up with Thingiverse for 3D printable designs.

Karan Chapekar, director, KCbots, says, "KCbots is collaborating with GreySim in order to make ultra-low 3D printers for students, to help them learn."

Rhishikesh Patankar, research analyst, Patent iNSIGHT Pro says, "Companies filing patents in India for 3D printing include Council of Scientific & Industrial Research, Cabot Corp, Disney Enterprises Inc., General Motors Corp and 3D MTP Ltd."

Patent wars?

Printing technologies with a twist

Certain 3D printing technologies have evolved with specific changes based on environment. Let us take a look at the most popular of these:

Carbon fibre. Carbon fibre can usually be incorporated into objects only by hand, so working with them is time consuming and proves expensive too. 3D printers use carbon fibre material in the form of composites, which cuts down waste by providing the exact physical length needed without any requirement to trim.

Light-directed electrophoretic deposition (EPD). EPD is an old technology that is unable to allow selective depositions at certain spots. A combination of photoconductive electrodes with a DC electric field allows spot deposition to produce accurate composites in a process called light-directed EPD. EPD being a faster deposition technique, the addition of light to it allows the creation of accurate void areas, thus helping the engineers to come up with all types of designs in a shorter span of time.

The interesting point in this technology is that it uses laser-cut aluminium mask, which is likely to be replaced soon by a digitally-projected mask that would make it a completely automated deposition system. Being a resin-type 3D printing system, it provides better finish to the object being printed.

Biomimicry. Biomimicry is a new science that studies nature's models and then uses these designs and processes to solve human problems. 3D printers can be a great boon in making products similar to those found in nature. The best

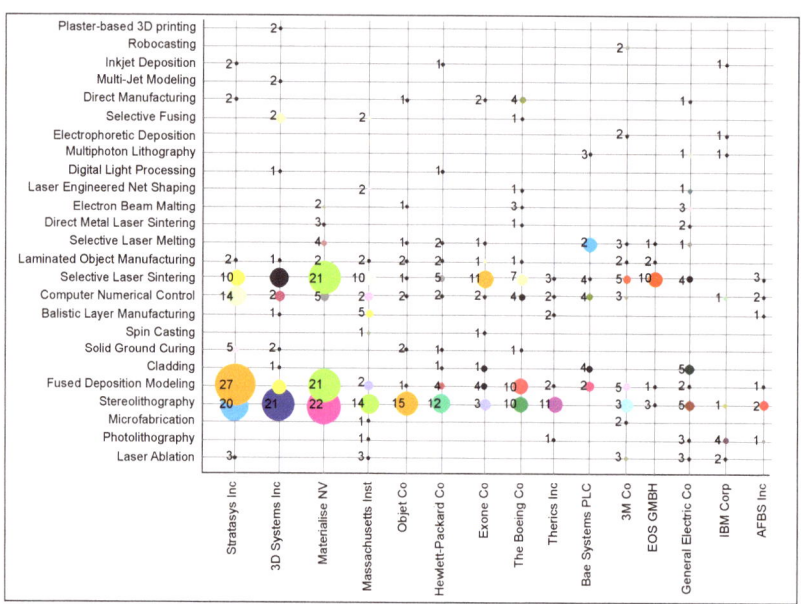

Fig. 1: Technologies used by top 15 brands (Courtesy: Patent iNSIGHT Pro)

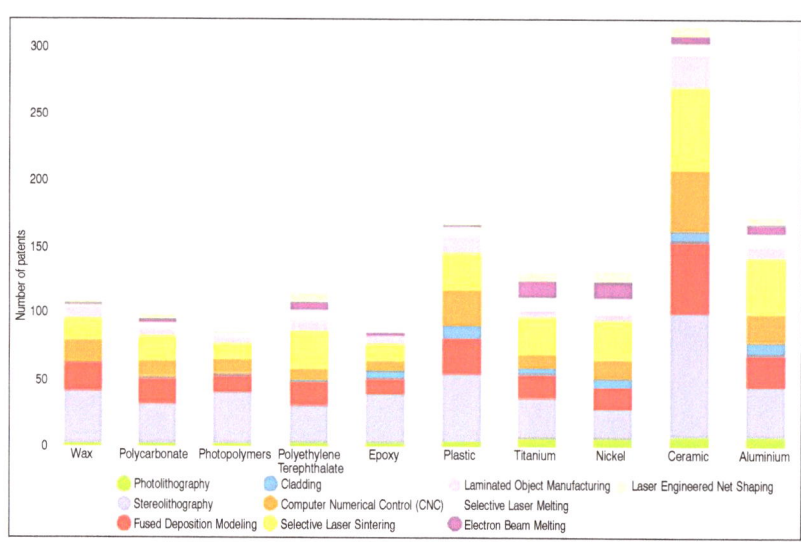

Fig. 2: Key materials used in different 3D printing technology (Courtesy: Patent iNSIGHT Pro)

thing about this technology is that the final products can get different colours without any extra cartridge attachment to the 3D printers. This lets us create new prototypes of different colours and shapes without any extra tools and cost.

Sensor and data analysis in 3D printing. Sensors are the key source for inspection in data analysis. They play an important role in improving the additive manufacturing process in 3D printing. If we take the example of selective layer sintering (SLS), the sensor would allow the engineers to identify the optimal coupling between the powders and the

laser by helping them to analyse the right amount of energy transferred into the powder layer. It helps to prevent overheating during the process.

New 3D printers in the market

The 3D printing industry can be seen as a growing industry with developments being made at every stage. No wonder newer technologies are emerging. But there is requirement of new 3D printers that can be used for multiple platforms with multiple filaments.

There is a 3D printer MarkOne, created by a start up named MarkForged,

STM32 L0
Ultra-low-power MCU

FOR ENERGY SENSITIVE APPLICATIONS

- 1.65 to 3.6 V ARM Cortex-M0+ based platform
- 16- to 192-Kbyte dual bank Flash, 20 Kbytes RAM, 6 Kbytes EEPROM
- 12-/16-bit ADC 1 Msps - 240 µA
- 139 µA/MHz full speed - 87 µA/MHz (optimized)
- 400 nA Stop mode + Full Ram (3.5 µs wakeup)
- -40°C to +125°C operating range

STMicroelectronics Pvt Ltd
Plot No 1, Knowledge Park III
Greater Noida 201308
Uttar Pradesh
India

For more information, visit www.st.com/stm32l0

Surprising innovations

There is some pretty impressive R&D going on in the world of 3D printing. Examples:

Living organs. Recently, a 3D elastic membrane has been developed using the 3D printing technology. This is a boon to the medical industry as it can help predict cardiac disorders and help the doctors correct them early. The speciality of this membrane is that it has tiny sensors printed on it, which provide detailed information even before the symptoms of the problem are exhibited by the patient physically.

Recent research has also seen the development of a 3D printed human liver and a heart, both made up of a transparent material and threaded with arteries and veins that are coloured. This is expected to help the physicians prevent the underlying complications during transplant.

3D printed edibles. It may sound crazy but it is a reality. A 3D printer called Foodini can create 3D-printed food with fresh ingredients, that too without the involvement of any cook. This printer can help create various shapes of the food that are generally difficult to create if you are not an expert. It can print various food items like chocolates, cookies of different shapes, hamburgers, etc. The only problem we see with this printer is that the food still needs to be cooked!

Green vehicles. Urbee is a 3D-printed car that is estimated to use about eight times less energy than an average car. It weighs less than 500kg and needs less than 38 litres of fuel to drive from San Francisco to New York. It marks an important landmark in the march towards green vehicles.

Ethical filament. Some 3D printing firms have partnered with the waste pickers in India to convert the waste plastic into filament for 3D printing. This would have both social as well as environmental impact on the society by empowering the waste pickers and eliminating the non-degradable plastic waste from the environment.

Major contributors to this report

➤ **Daniel Hutchison,** co-founder, HYREL LLC
➤ **Harshad Karmarkar,** research analyst, Patent iNSIGHT Pro
➤ **Karan Chapekar,** director, KCbots
➤ **P. Rajasekar,** Accurate 3D
➤ **Paul Anand,** CEO, Biotz
➤ **Priya Kuber,** managing director, Arduino India
➤ **Rhishikesh Patankar,** research analyst, Patent iNSIGHT Pro

at Boston, which can print carbon fibre. As mentioned earlier, though carbon fibre can be used to make things manually, it is tough to work with and time consuming as well. MarkOne can make working with carbon fibres easy and less time consuming, without the requirement to trim the object as only the required amount of carbon fibre is laid down. Thus a strong object can be created in less time with least effort. This printer not only works with carbon fibre but also has the ability to work with other types of composites like nylon, fiberglass and PLA (polylactic acid).

Then there is anther printer called Duplicator 4, which is an assembled dual-extrusion 3D printer available in India. Although it is capable enough to be compared with MakerBot's Replicator 2x, it is almost one third in price. It can 3D print in two colours and also has the capability to use one extruder to print the actual object and the other to print the dissolvable material. It uses the FDM technology and can use both plastics and composites for the filament. Karan says, "Duplicator 4 is a blatant copy, a clone or a duplicate of MakerBot design made in China and imported by some people in India."

Usually, the printers are not something you can assemble at home, but with 3D printers this is not the case. Besides buying a low-cost 3D printer you also have the option to assemble your own 3D printer. There are many online global communities like RepRap that offer the do-it-yourself kits for assembling your own 3D printer.

Daniel Hutchison, co-founder, HYREL LLC says, "We have come up with Mk1 filament extruder and EMO-25 extruder, where EMO-25 can print precious-metal clay on all platforms and uses most of the emulsified filaments like PMC and silicon. "Biotz is introducing a disruptive portable 3D printer which would be affordable for each and every home by 2015," says Paul.

Want to try 3D printing?

New technologies require new tools, software and training to support it. Some companies like Redd Robotics, Cycloid System and Kurukshetra are organising workshops in India to help people understand the concept of 3D printing and to encourage this technology for the growth of the Indian market. Leading firm like Ultimatum and SAP Labs are working with experts in this industry to explore new opportunities for on-demand development of new technologies and software anywhere and anytime.

The software plays an important role in the designing and development before actual printing takes place. The 3D models designed are first verified and tested to ensure that the final product to be printed would meet the requirements of the customers. Karan says, "The major supplier of 3D printing software is Autodesk with their key software being AutoCAD. They are also releasing a number of free software related to 3D printing, one such example being 123D Design that is used for 3D modelling."

Then there are other software like Meshmixer that allows the designers to mix and match, stamp, script and paint one's own 3D printed designs. The user-friendly online tool like TinkerCAD allows the designers to convert their ideas into a CAD models for 3D printing. Harshad Karmarkar, research analyst, Patent iNSIGHT Pro says, "Although CAD/CAM software forms a base for 3D printing, companies have developed their own printers for specific applications. For example, 3D Systems has come up with a printer which prints on pizza."

Varying price

The main focus of the Indian market is low-priced 3D printers. Even though the consumers are apprehensive about quality, the main concern of the market is to ensure that the low-priced 3D printers meet the customers' requirement. Daniel Hutchison says, "The current price trends that we can see in the FDM or FFF market have been more towards making the cheapest printer possible for the consumer-level marketplace. SL has been price trending the same way although SLS is still an enigma at the moment." He opines, "It will be interesting to see how HP and other large companies would affect the 3D printing marketplace when they make their entrance in the near future." ●

The author is a technology journalist at EFY

THE FIRST LINE OF DEFENSE BEGINS AT THE CHIP LEVEL

Today's interconnected factories face security threats from every corner. Don't wait for your customers to demand security. Be one step ahead. DeepCover® solutions offer strong IC-level protection, so you can safeguard your systems and help avoid millions in losses. *maximintegrated.com/deepcover-security*

3D Printers to Print Lunar Base and Living Tissue

From being used to prototype designs and build custom toys, 3D printing technology is now ready to churn out everything from space stations and human organs to houses and food too

JANANI GOPALAKRISHNAN VIKRAM

The International CES is an indicator of promising technologies, and this year 3D printing was an extremely popular theme, along with the Internet of Things, wearable devices, 3D televisions and other emerging tech. As Avi Reichental, president and CEO of 3D Systems wisely pointed out at the event, "3D printing is an overnight success 30 years in the making." But then, it is true that a few years ago MakerBot was perhaps the only 3D printing player at CES. Today, there are several players vying to capture this growing market, including XYZ Printing, 3D Systems, MakerBot and more.

It is also interesting to note that these printers are getting more and more versatile and cost-effective too. There are sub-$500 models such as Printbot Simple and Da Vinci 3D Printer. There are those like Replicator Mini, Solidoodle and Cube3, which are priced quite effectively for their capabilities. And, there are also innovative, awe-inspiring ones like the ChefJet 3D Pro that can print edible food models in full colour!

The trend is rather strong, and it is obvious that today it is technically possible to create anything from food and toys to weapons and prosthetics using these 3D printers. While currently 3D printers have not captured the fancy of home users due to the lack of simple and user-friendly software tools, a research report released this April by Juniper Research says that sales of 3D printers for home-use will exceed one million units by 2018, thanks to killer

Multi-dome lunar base built by the European Space Agency using 3D printing techniques (Courtesy: ESA/ Foster+Partners)

apps, expansion of use cases and entry of established players like HP.

The industry and academia are also in tune with this trend, and are gearing up with improved technologies, materials and processes to take 3D printing to greater heights. Here is a sample of some such R&D activity in the 3D printing space and innovative applications of the same.

Building dream homes. Kamermaker, which means 'room builder,' is an open source 3D printing pavilion developed by Ultimaker. The six metres tall printer builds large plastic-based building blocks that can be used in construction. And these building blocks are nothing like the rectangular bricks used in construction. They are large chunks of the building itself—say a quarter of a room or a staircase—which just need to be patched together to make the house.

A Dutch architectural firm is now working on building a canal house in Amsterdam using the Kamermaker. The

project is apparently an experiment that applies 3D printing to construction, and so the architects have left the construction site open to public viewing. They believe that they might have to re-build parts of the house several times till they arrive at the perfect design. This demonstrates the flexibility that 3D printing offers. The canal house might mark the beginning of an era where people can buy architectural designs off stores and have their house printed within weeks.

Space stations too. Last year, the European Space Agency (ESA) teamed up with architects to test the use of 3D printing for more efficient lunar base construction using materials like moon dust. The lunar base, designed by Foster+Partners, has a multi-dome design with a cellular structured wall as protection against micrometeoroids and space radiation. It has inflatable structures to accommodate astronauts. The lunar base is being 3D printed using a special printer supplied by

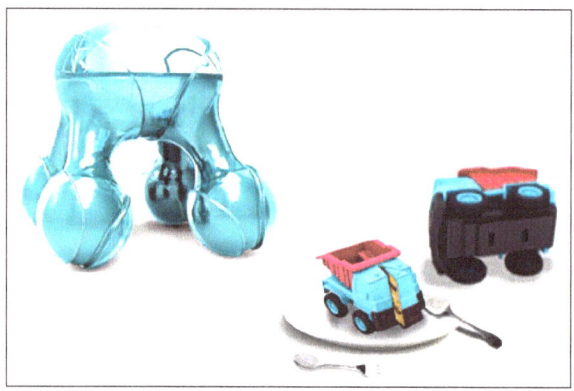

Atomium prints food in designs that appeal to kids (Courtesy: Luiza Silva and Electrolux Design Labs)

UK-based Monolite. As a demonstration, they have printed a 1.5-tonne building block using lunar soil. The block has a hollow closed-cell structure inspired by bird bones, to provide the right balance of strength and weight.

Lord Brahma has competition. Bioprinters are capable of 3D printing human tissue too. Using bio-ink made of living cell mixtures, human tissue can be printed layer by layer. Layers of hydrogel applied between layers of cells work as a scaffolding. Once the cells fuse, the hydrogel is removed. This structure is then transferred to a bioreactor for Nature to work her magic. The tissue grows naturally into its final form.

Today, scientists are able to print tissues such as skin and cartilage but there is hope that life-saving organs will be printed in the future. In a recent breakthrough, Harvard researchers have applied micro-scale 3D printing to create tissue containing skin cells and biological structural material interwoven with blood-vessel-like structures. Using a customised four-head 3D printer, they have created hollow, tube-like structures within a mesh of printed cells using special ink that liquefies as it cools.

This is considered a disruptive innovation because although several scientists have had success in building organs using varied techniques in the past, they have always been stumped by the lack of blood vessels. This recent development achieves precisely that, using 3D printing!

Picture perfect. One of the winning inventions at this year's Innovation Days at the University of Wisconsin-Madison was Spectrom, an attachment for 3D printers that enables users to incorporate 'seamless, on-demand colour' into the 3D printing process. Developed by two chemical engineers, Spectrom is designed to work with fused deposition modelling (FDM) 3D printers, which are the most common amongst consumer printers due to their ease of use and relatively low cost.

Current FDM colour printing technology simply makes use of pre-dyed plastic filament, limiting opportunities for the user. Spectrom on the other hand delivers solvent dyes directly to clear plastic filament in a continuous, on-demand process. This allows the user to print colourful 3D models easily, just as we print colour pictures using an inkjet printer. The inventors think that colour-matched prosthetics is one of the most useful applications of Spectrom.

Currently, doctors mould parts such as noses and then have an artist paint them to match the patients' skin tones. But, with Spectrom and a 3D printer, it is possible to scan someone's face, build an exact face profile and easily print perfectly-matched parts.

3D printed skull helps patient regain vision and normal activity. Early this year, a team of Dutch surgeons led by Dr Bon Verweij successfully replaced the skull of a 22-year old with a custom-made, plastic, 3D printed skull. After a three-month observation, the doctors announced the success of the surgery in March. The 3D printing technology used was developed by Anatomics of Australia. The researchers said that similar surgical techniques could be used to treat several other bone abnormalities.

Beating hunger pangs, even in outer space. Several food printing solutions are in the market today, to print out food structures using raw material in powdered form. Today, most of them are busy churning out fancy stuff using raw materials such as cocoa powder and sugar. Like the figurine atop a wedding cake for example! However, there are also certain models that take protein powders or other nutrient molecules as input and print out balanced diets. Electrolux Design Labs' Atomium 3D Food Printer, starts with nutrient molecules, and fuses them layer by layer to create customised nutritious foods in various shapes. The printer will apparently solve the problem of serving healthy food to kids.

NASA too is working on a solution to 3D print nutritious food for astronauts in space.

US start-up Modern Meadow is working on a 3D printer that is capable of printing raw meat. The process starts with stem cells from the animals. The cells are then made to multiply and, once they reach large enough numbers, they are placed in a cartridge to act as bio-ink. The ink is squirted out of a nozzle to create desired shapes, after which the bio-ink particles naturally fuse to form real living tissue similar to natural raw meat. The team is working to address quality and food safety regulations. Once the technology is ready for the market, it would consume 96 per cent less water and 45 per cent less energy than the meat production methods prevalent today, and also reduce greenhouse gas emissions from this sector by 96 per cent, according to the company.

This obviously is just the tip of the iceberg. With ongoing innovations in electronics and material sciences, newer 3D printing technologies and inks for the same are cropping up, making it possible to print not just models but real-life objects and even living tissues. What we still hope for is more user-friendly software to design and print objects so 3D printing will catch up amongst home users as well, rather than remaining the preserve of scientists, technologists and expert do-it-yourself enthusiasts. ●

The author is a technically-qualified freelance writer, editor and hands-on mom based in Chennai

True low power LCD MCUs with sub μA display driver! Scalable for various LCD applications for healthcare, appliances, meters and more

RL78/L1C and RL78/L13 Group microcontrollers, members of the renowned 16-bit RL78 family, feature LCD panel drivers with ultra-low consumption LCD circuits. This makes them ideal for LCD display functions in home appliances (IH rice cookers, microwave ovens, water heaters, etc.), healthcare devices (sphygmomanometers, body fat scales, pedometers, etc.), measuring equipment and other products requiring high energy efficiency

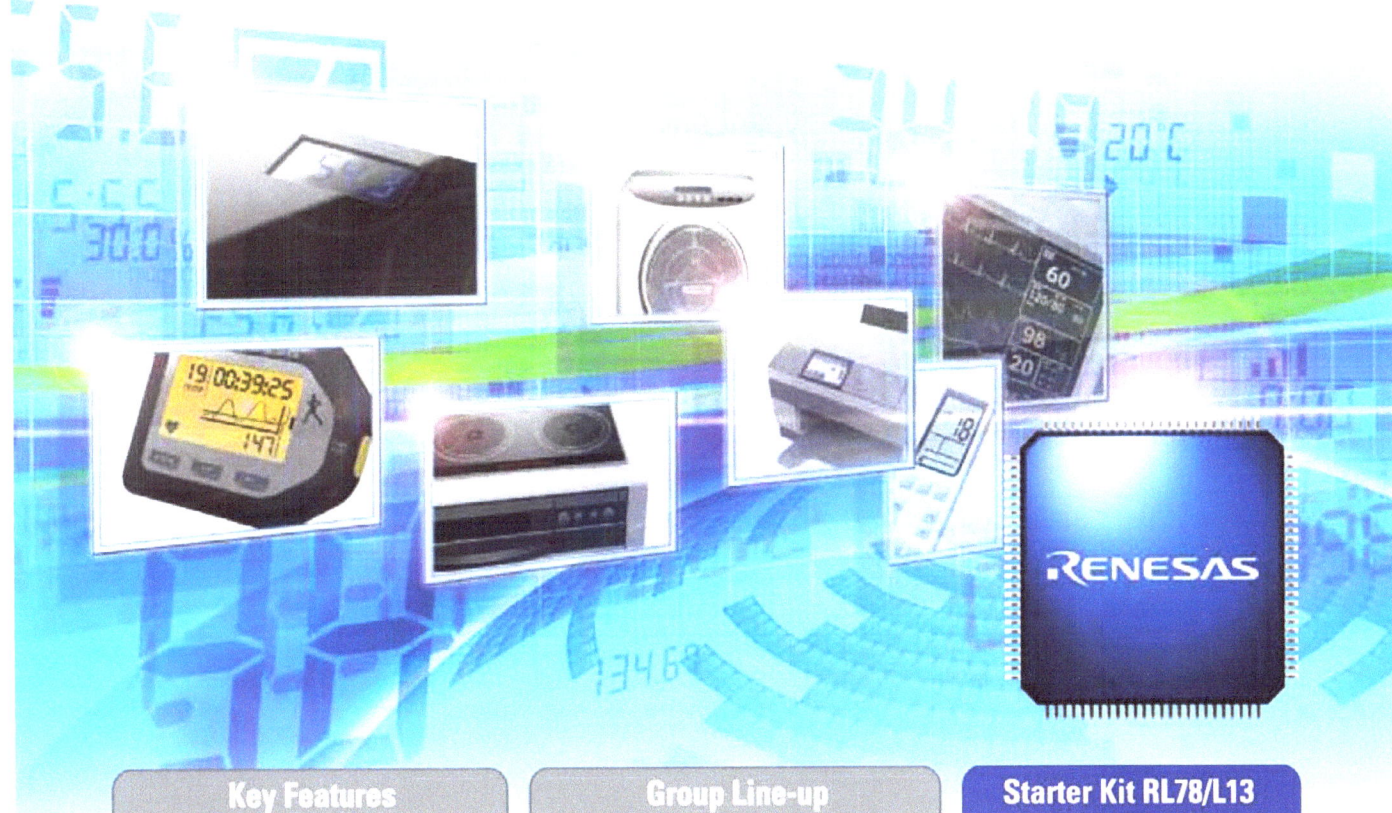

Key Features

- High-precision (±1%) on-chip oscillator enables 24 MHz CPU operation

- Ultra-low consumption LCD circuit

- Support for three methods of generating LCD drive voltage (internal voltage boost, capacitor split, external resistance division)

- Support for 18-step LCD contrast adjustment

- Support for LCD blinking

Group Line-up

◢ RL78/L1C main features
- Support up to 416 LCD segments, (44 to 56 seg X 4 com or 40 to 52 seg x 8 com)
- 12-bit ADC / 8-bit DAC
- USB 2.0 function
- Battery charging 1.2 support
- Packages: 80 an 100-pin packages

◢ RL78/L13 main features
- Support for up to 240 LCD segments (51 seg x 4 com or 47 seg x 8 com)
- ROM capacities: 16 to 128 KB
- Packages: 64 and 80-pin packages

Starter Kit RL78/L13

Explore now!

http://in.renesas.com/rsk

Renesas Electronics India

 Join us on Facebook

http://in.renesas.com

RENESAS

Machine-Brain Interface: A Giant Step for Mankind

As the brain-computer interface is being closely coupled with bionics, the medical field is witnessing a breakthrough, giving hopes to millions of amputated and paralysed people around the globe

ANAGHA P.

Jan Scheurmann was paralysed from neck down more than a decade ago due to degenerative brain disease. But thanks to the brain-computer interface (BCI), now she could grab and eat a chocolate with her robotic arm and even hi-five her doctors! With two aspirin-sized electrodes implanted to her brain and a few months of training, she was able to successfully manipulate a mind-controlled robot arm with seven axes of movement (front-back, up-down, left-right, wrist yaw, wrist pitch, wrist roll and hand grasp). After grabbing her first bite of chocolate she declared, "One small nibble for a woman, one giant bite for BCI."

"This is a spectacular leap towards greater function and independence for people who are unable to move their own arms," the senior investigator of the University of Pittsburgh's Pitt School of Medicine, Andrew Schwartz, said in a release. "This technology, which interprets brain signals to guide a robot arm, has enormous potential that we are continuing to explore. Our study has shown us that it is technically feasible to restore ability; the participants have told us that BCI gives them hope for the future."

In an earlier independent study by the BrainGate Implant, another paralysed woman, Cathy Hutchinson, was able to use the arm to pick up a thermos of coffee and drink it from a straw.

What is BCI

BCI, also known as a brain–machine interface (BMI), mind-machine inter-

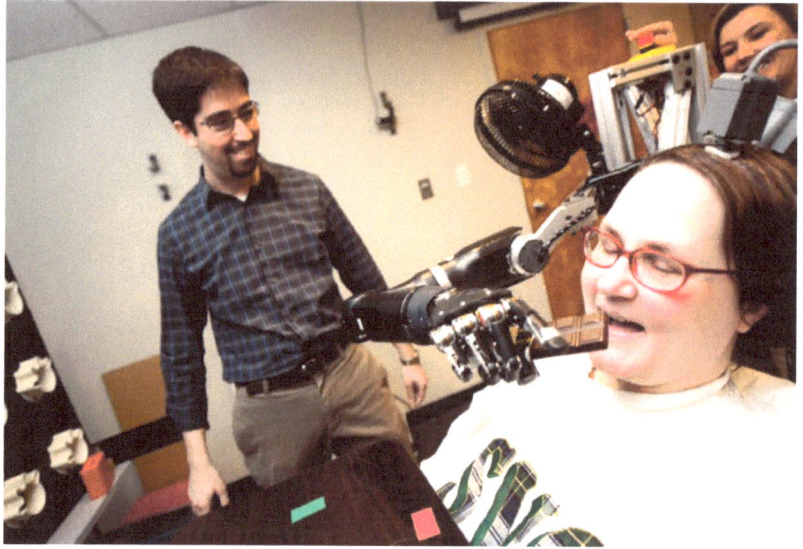

Jan Scheurmann (Courtesy: http://www.upmc.com)

face (MMI) or direct neural interface, is a method by which the brain signals are used to control an external device. In simple terms, this technology helps you to control machines with thoughts. BCI has become one of the important areas of research in medical field as it can restore and augment human sensory, motor and cognitive functions.

The BCI system

A long-term goal of the BCI projects is to develop a neurotechnology that could convert thoughts into actions, thereby helping people with limb loss or paralysis to restore their movement and control, making them independent to a great extent.

The current BCI system comprises three main segments:

Sensor. A device implanted in the brain, usually multi-electrode arrays (MEAs), that records signals directly

related to imagined limb movement.

Decoder. A set of computers and specialised programs that can interpret the neural signals collected by the sensor, and which turn them to corresponding commands for an external device.

External device. A communication device like computer or a robotic limb on which the decoded brain signals are converted to required actions.

How it works

The first step is to associate raw neuronal signals with their corresponding movements. The magnetic resonance imaging mapping is used to find out the regions of the brain that are active when performing certain functions. This helps to place the electrode in the right target.

The sensors are smaller than a square centimetre and are implanted in the primary cortex of the brain. Each

Do you speak MATLAB?

Over one million people around the
world speak MATLAB.
Engineers and scientists in every field
from aerospace and semiconductors
to biotech, financial services, and
earth and ocean sciences use it to
express their ideas.
Do you speak MATLAB?

*Saturn's northern latitudes
and the moon Mimas.
Image from the
Cassini-Huygens mission.*

*Related article at
mathworks.in/india*

MATLAB®
The **language** of technical computing

of these MEAs has around hundred contact points that record the response of individual neurons as well as populations of neurons.

The microchips are first implanted in brain. The patient is made to think about doing something, like moving the hand forward or rolling the wrist. Each activity triggers a unique set of neurons. When a neuron is excited, a potential difference is developed between the inside and outside of the cell membrane and an ion current is developed. This ion current is converted into electron current by the sensors and sent to the decoder.

The difficult task is to interpret and associate the brain signals to their respective movements. With the help of revised filtering, processing and complex computer programs, and after many trials, the signals related to each activity can be figured out. The specially designed embedded software is programmed so as to receive each signal and make the connected external device (like a prosthetic hand) physically perform the activity associated with that particular signal.

Commercial viability

Though extensive researches are going on in this field, only a very small number of products are available in the consumer market; even lesser belonging to medical industry. High cost of production and unavailability of participants for clinical trials are a few reasons for their reduced popularity.

Some of the medical devices based on neurotechnology are discussed in brief below:

BrainGate implant. It is a brain implant developed by Cyberkinetics and is one of the most important BCI systems.

Deep brain stimulator. It is manufactured by Medtronic and is used to treat Parkinson's and dystonia. It is also researched for treatment of epilepsy and depression.

Hearing aid. The medical device company Otologics has made this product available in Europe.

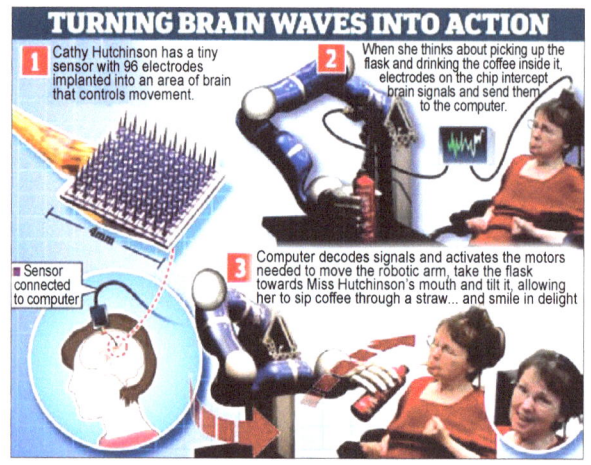

Cathy Hutchinson (Courtesy: http://www.dailymail.co.uk)

Bionic eye. An artificial retina by Second Sight Medical Products, going through the first phase of trial.

Vagus nerve stimulator. It is used for treating epilepsy, depression and obesity. Cyberonics, MetaCure and EnteroMedics are the major manufacturers.

Powered exoskeleton. A robotic exoskeleton that can help the elderly and the disabled to walk and lift objects, developed by Cyberdyne.

Implanted defibrillator. This device is meant for people who have problems with irregular heartbeat. It prevents sudden cardiac arrest. It is developed by Medtronic, St. Jude Medical, Boston Scientific, Sorin Group CRM and Biotronic.

Ventricular assist. It helps the heart chamber pump blood. Thoratec, Abiomed, Inc. and World Health Corporation manufacture this product.

Breathing pacemaker. It helps people with spinal cord injury to help them breathe on their own by pacing diaphragm. The device is produced by Synapse Biomedical, Inc.

PillCam. A capsule with miniature camera that can be swallowed and is used for endoscopy. Given Imaging is the manufacturer of this capsule.

Drug delivery to spine. This drug delivery system is meant for patients with chronic pain. It is produced by Medtronic and MicroCHIPS.

Artificial arm. Prosthetic arms with 18 degrees of freedom, developed by

DEKA Research and Development Corporation.

Bionic hand. This product designed by Touch Bionics can perform more hand functions than a normal prosthetic hand.

Sacral nerve stimulation. Developed by Medtronic to control some bladder functions.

Smart leg. Most advanced lower limb prosthetic by Otto Bock Healthcare.

Robotic foot. It has many added features over prosthetic foot. Ossur is the manufacturer of this product.

What future holds

As the next step of this technological advancement, researchers are planning to introduce feedback potentials to the electrodes, which can result in the interpretation of sensations like grip strength. They are also working on miniaturisation of equipment and introduction of wireless technology. It may even be possible to avoid the robotic devices by directly bypassing the appropriate signals to the corresponding motor nerves in the damaged section of the spinal cord. This means the paralysed patient will be able to move his/her own body.

The field of BMI has made a considerable development in the past decade. Though only few products are available in the market, extensive research and lab works are going on in this field. What was once just an interesting science fiction theme is now growing very close to reality. Neurotechnology has developed from just a figment of imagination to monkeys moving cursors on computer screen to humans controlling prosthetic limbs in three-dimensional space with dexterity. In coming years, we can expect BCI to not only restore human functions but also augment it and improve the quality of living. ●

The author is a technical correspondent at EFY

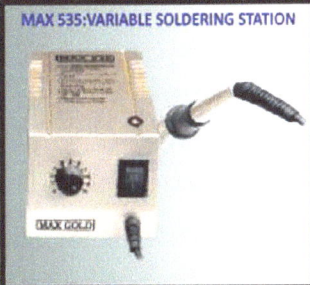

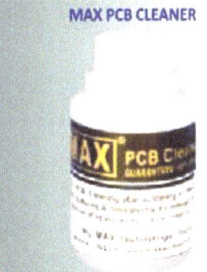
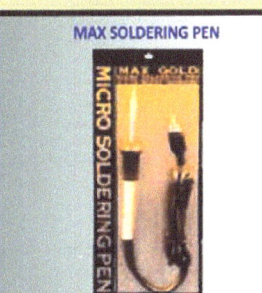

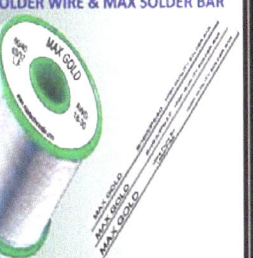

Smart Glasses Just Got Smarter

Golden-i Gen 3.8 is a handsfree wireless computing and communications headset that operates by voice commands and head movements. It was demonstrated at Electronics Rocks Conference 2013

ANAGHA P.

We all have seen the media frenzy that surrounded Google's Glass, a wearable computer that integrated a lot of Android functionality, while making you look like an Android at the same time.

Golden-i is a mobile wireless wearable headset computer with 38cm (15-inch) near-eye virtual display, which enables the users to carry out general computer functions without using their hands. It allows the users to access six independent devices or networks simultaneously (including PCs, cellular phones and industrial and enterprise servers). The device supports more than 38 languages. It is operated by voice-user interface and head-tracking system.

Technology

The device has an embedded 14MP camera that can record and/or send full-colour images or streaming video in real time, in resolutions ranging from 240p to 1080p, at the rate of 2fps to 60fps. In addition, it is equipped with a head tracker, Wi-Fi, Bluetooth, ambient-light sensor, USB on-the-go and other peripherals. The design also includes a low-power, ultra-small form factor micro-display that integrates advanced voice and gesture recognition technology.

The latest version of Golden-i, the Gen 3.8 system, is built around TI's powerful OMAP4 processor and runs on Android 4.2 operating system. A Windows Embedded Compact 7 version of the product is also supported. An automated test suite is available for hardware design validation and manufacturing test.

Construction worker using Golden-i

Applications

Golden-i is a finished product and is being adopted in various fields, such as fire brigade, police department, construction, maintenance, repair and general professional applications. It is used in healthcare by doctors and nurses, and for emergency medical services. Product developers can also use Golden-i as a development platform.

Since this gadget is highly customisable, it can be used in a variety of applications, and robust solutions and support are provided to the clients by the firm.

Here is an example of how Golden-i Gen 3.8 is used in a fire brigade. The application used here is the Firefighter Pro, which helps improve time efficiency and safety. It allows firefighters to access vital information while dealing with an emergency situation. They could call up floor plans and GPS coordinates on their wearable computer, giving them real-time understanding of the layout, assisting navigation through unknown structures efficiently (using assisted-GPS inertial navigation) and hence enabling quick evacuation/rescue operations. It helps the firefighter see through smoke, mist, dust and thin materials using an optional IR camera.

The firefighters can monitor the vital signs of individuals and the crew (which is an application in healthcare field too). Live, on-site video streaming is also possible to and from the affected environment.

Development

The Golden-i is a global multinational endeavour that was developed with the goal of improving productivity, efficiency and safety in all professional and industrial sectors. It supports and guides the user at any moment without interruption.

The US-based Kopin Corporation's portfolio includes designing, building and deploying lightweight, power-efficient, ultra-small liquid crystal displays. They entered into a partnership with Bangalore-based Mistral Solutions to design the headset computing and communications system. Mistral also manufactures compact hardware electronics required by Golden-i.

The concept of handsfree, wireless, headset computer was first discussed in August 2006. The first actual Golden-i headset development started on July 1, 2007, and in January 2008, the first handsfree computing wireless Golden-i headset was demonstrated streaming 480p video at 25fps over Bluetooth 2.0 EDR. It used a speech recognition user interface with approximately 85 per cent efficiency.

Kopin Corporation has over 200 issued patents covering various aspects of Golden-i. Apart from this, the company also has an additional 100+ patent disclosures open, and approximately 60 additional patents filed since 2007, which are being processed.

The team is currently working with the world's leading cellular service providers (such as Verizon Wireless) for bringing the Golden-i Gen 3.8 headsets in various global geographical markets this year.

Design challenges

"The major challenge while developing this product was fitting all the interfaces into a small form factor design," according to Selvaraj Kali-yappan, general manager - Delivery (PES), Mistral Solutions. The product should offer multiple technologies, such as cameras (infrared, SWIR, near-IR, visible light and terahertz), Wi-Fi, Bluetooth, 4G LTE cellular, 9-axis head tracking and numerous sensors while also ensuring sleek, lightweight and compact design. Space optimisation was achieved by using microBGA and package-on-package (PoP) technology. The Mistral team also addressed multiple routing and layout challenges.

The overall power consumption and heat dissipation in the system was also successfully reduced, thus ensuring longer duration for system operation.

Since Golden-i relies much on voice commands, implementation of a stable voice recognition and active noise cancellation in ambient environments with as much as 120dB of noise was yet another challenge. The algorithm integration was to be performed to the desired level to meet the specific market demographics and ambient environments that are typically encountered.

Best of its kind

Google Glass is a similar product that made huge headlines last year. But original Golden-i had reached the market ahead of Glass.

"Golden-i is not a single device, but a family of devices leveraging similar hardware and software," says Jeff Jacobsen, senior advisor to CEO, Kopin Corporation, "but significantly differing in headset size and peripherals, depending on the market being addressed."

The initial version of Golden-i 3.8 had been designed for use with hard hats and helmets, for industrial use and applications. Smaller, sleeker and more stylish models of Golden-i are expected to hit the market by 2015. ●

The author is a technical correspondent at EFY

How to Protect Telecom Networks

Automatic protection switching schemes allow configuring a pair of lines for line redundancy. When the working interface fails, the protection interface quickly assumes the network traffic and provides end users with a continuous operation. Let's delve deep into protection mechanisms...

DR RAJIV KUMAR SINGH

Network failures, whether due to human error or faulty technology, can be very expensive for users and telecom service providers alike. As a result, the subject of so-called fall-back mechanism is currently one of the most talked about in the telecom world. A wide range of standardised mechanisms is incorporated into synchronous networks in order to compensate for failures in network path/elements and to provide highly-available telecom networks.

Automatic protection switching

Automatic protection switching (APS) is a fault-tolerant topology that is used for providing backup to telecom networks. For network survivability, in the event of failure in a network element or link, APS involves reserving a protection channel with the same capacity as the channel or facility to be protected. In the event of signal-fail (SF) or signal-degrade (SD) condition, the working line switches automatically to the protection line within a few milliseconds.

Fig. 1 shows basic APS configuration on network elements, for example, routers/nodes A and B. Here, node A is configured with the working interface and node B is configured with protection interface. In a router configured for APS, configuration for the protection interface includes the IP address of the router (normally its loopback address) that has the working interface. Normally, the working and protection interfaces are connected to a network element of the transmission system, typically, an add-drop multi-plexer (ADM).

In the event of failure on working interface of node A, the connection automatically switches over to the protection interface on node B. On the protection circuit, K1 and K2 bytes from the line overhead (LOH) of the

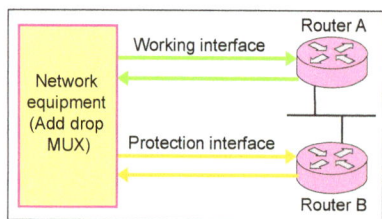

Fig. 1: Basic APS architecture

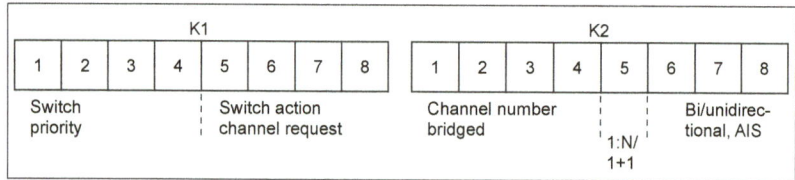

Fig. 2: The K1, K2 bytes

synchronous digital hierarchy (SDH) frame indicate the current status of the APS connection and convey any requests for action.

K1 and K2 bytes. The K1 byte in SDH configuration contains switching pre-emption priorities (in bits 1 to 4) and channel number of the channel requesting action (in bits 5 to 8). The K2 byte contains channel number of the channel that is bridged onto protection (bits 1 to 4) and mode type (bit 5); besides, bits 6 to 8 contain various conditions, such as multiplex-section alarm-indication signal (MS-AIS) and indication of unidirectional or bidirectional switching (Fig. 2).

The APS is very extensible in terms of topologies (for example, rings) and flexibility (for example, link-capacity-

adjustment scheme (LCAS) service restoration). Basically, two types of protection architectures—linear protection and ring protection—are distinguished in APS. The linear-protection mechanism is adopted for point-to-point connections. But ring-protection mechanism can take on many different forms. Both mechanisms use spare circuits or components to provide the back-up path.

Linear protection

1+1 APS architecture. The simplest form of mechanism for network survivability in the event of network failure is 1+1 APS. Here, each and every working transmission path/line/channel is protected by one protection path/line/channel (Fig. 3). At the near end, the signal is bridged permanently, that is, split into two identical signals, and sent over both the working and the protection lines simultaneously. At the far end, signal selection is made on the basis of switch initiation/trigger criteria, which are signal fail (SF), signal degrade (SD), loss of signal (LOS) or loss of frame (LOF).

If a defect occurs, the protection agent/switch in the network elements at both ends switches the circuit over to the protection line. Switching at the far end is initiated by the return of an acknowledgment in the backward channel. 1+1 architecture includes

Test Probes

for ICT and Cable Test

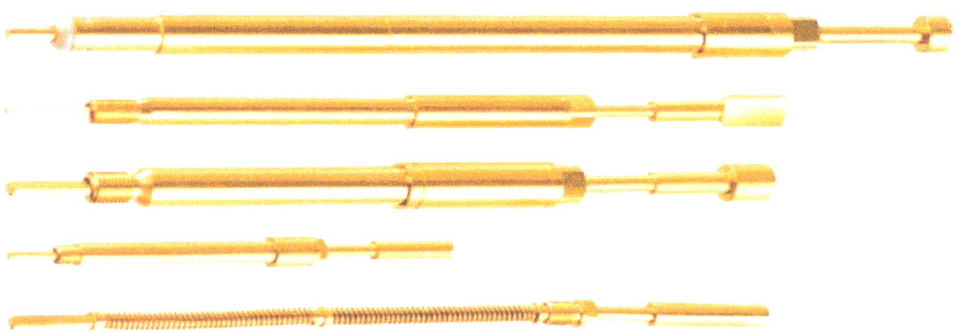

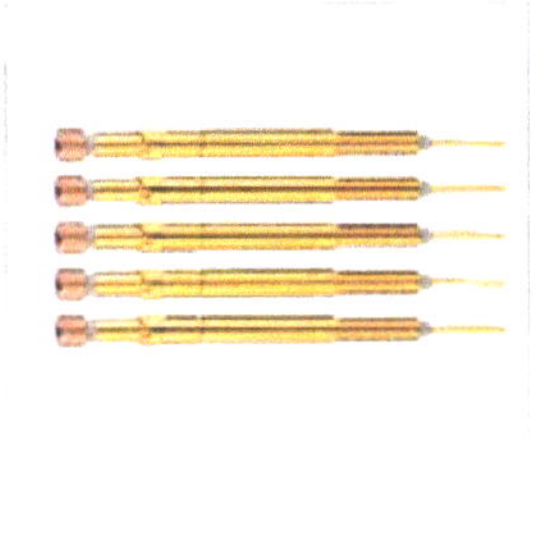

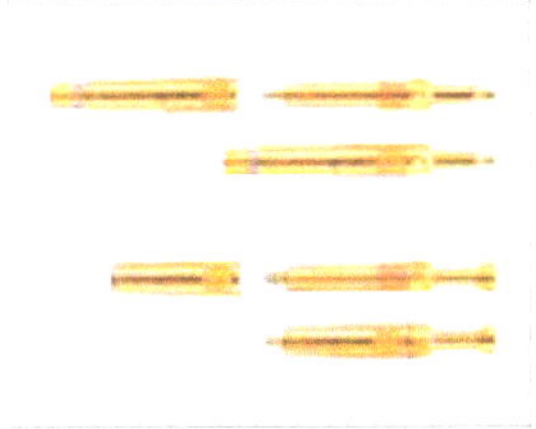

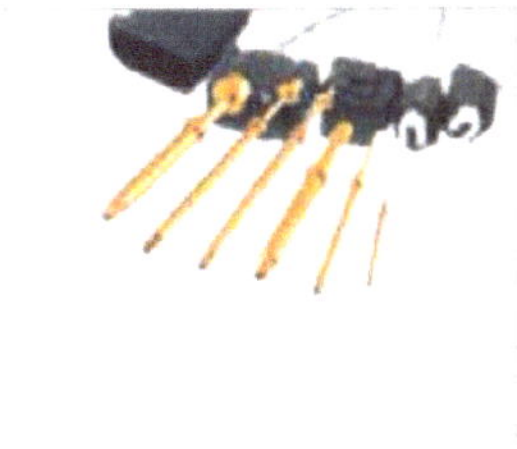

Innovative Products

PTR offers a wide rang of test probes for all applications: Switching test probes and test probes with thread for the harness test, test probes for testing printed circuits (ICT and FT), battery probes and interface contacts, high current test probes for run-in or burn-intest, pneumatic test probes for functional test application.

 MADHU SUBTRONIC COMPONENTS PVT. LTD.

Since 1976

Contact Person: Mr. Pratik Lakhani
1st Floor, Krishna Building, Shamrao Vithal Marg, Off. Lamington Road, Mumbai - 400 007, India
Mob :- +91-9821617949   Phone :- 022 2387 6707 / 022 2380 0783, Fax :- 022 2387 6729

P T R

A Phoenix Mecano Company

Contact Us : cs@mscpl.com

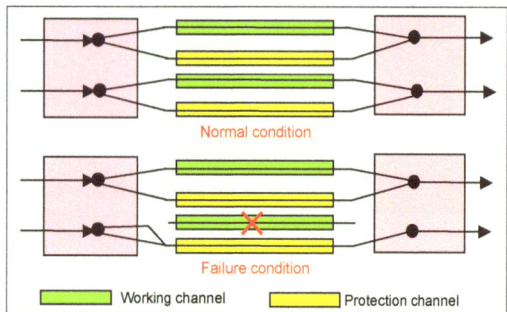

Fig. 3: 1+1 linear protection switching

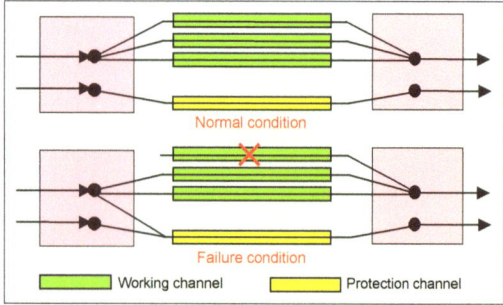

Fig. 4: 1:N linear protection switching

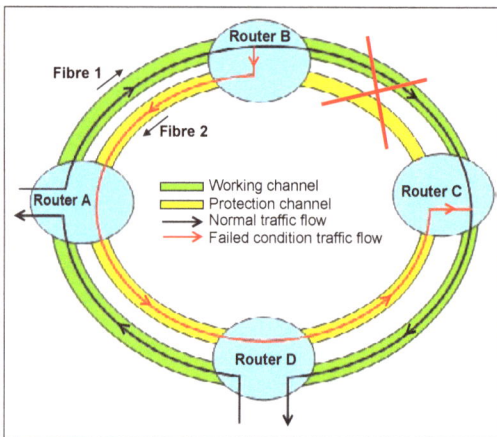

Fig. 5: Unidirectional ring protection

switched over to the back-up line/channel. During normal operation, no traffic or low-priority traffic is sent through the protection/redundant path.

When any failure occurs (such as, fibre-cut), both the source and destination switch onto the redundant or alternate path. Here, all switching is revertive, which means, the traffic shifts to the working line as soon as the failure is corrected. The reserve circuits can be used for lower-priority traffic, which is simply interrupted if the circuit is needed to replace a failed working line. Although network utilisation is better in this architecture, it requires signalling overhead and also results in slower restoration.

Ring protection

A ring is the simplest and most cost-effective way of linking a number of network elements. The greater the communication bandwidth carried by transmission media, the greater the cost advantages of ring structures as compared to linear structures. Various protection mechanisms are available for ring architecture, such as, unidirectional, bidirectional, revertive and non-revertive connections.

Unidirectional mode means that the two network elements (NEs) choose independently which circuit to receive, without negotiation. In all the modes, the working and protection interfaces receive the same payload from add-drop multiplexer (ADM), but only one is selected or currently active. The deselected interface is held in a 'line protocol is down' state and is completely removed. Only the selected interface actually processes the payload.

In bidirectional mode, receive and transmit channels are switched as a pair. But transmit and receive channels are switched independently in the unidirectional mode. For example, in bidirectional mode, if receive channel on the working interface has a failure event, both transmit and receive channels are switched.

In revertive connection, the hardware switches back to the working line automatically after repair of the working line or after the elapse of a configured period. In the non-revertive connection, if a failure condition occurs, the hardware switches to the protection line and does not automatically revert to the working line.

Unidirectional rings. In ring topology, traffic is transmitted simultaneously over the working and the protection lines. If there is an interruption, the receiver switches to the protection line and immediately takes up the connection. This switching process is referred to as line switching. A simpler method is to use the so-called path-switching ring in which a backup path is used from the source to its destination to bypass the failure.

Fig. 5 shows the basic principle of APS for unidirectional rings. Let us assume that there is an interruption in the circuit between the network elements, say, router B and C. In this situation, node adjacent to the fault will detect the condition and start the APS protocol. K1 and K2 bytes of the SDH frame indicate the current status of the APS connection and convey bridge requests, node information, type of failure, etc to the affected nodes. Each node detecting a fault sends an APS request to the node to which it was connected in the direction of fault. The connection is therefore switched to the alternative path in network elements.

Bidirectional rings. In this network architecture, connections between network elements are bidirectional. Often, bidirectional line-switched ring (BLSR) is used in which the overall capacity of the network can be split up for several paths, each with one bidirectional working line. For unidirectional rings,

100 per cent redundancy, as there is a spare line for each working line. This architecture is simple for implementation and results in fast restoration. But, its major drawback is the wastage of bandwidth, since no useful traffic travels through the redundant paths.

1:N APS architecture. Economic considerations have led to the preferential use of 1:N architecture, particularly for long-distance paths. In this case, a single back-up line protects several working lines (Fig. 4). When the primary path/channel fails, the two ends of the affected path are

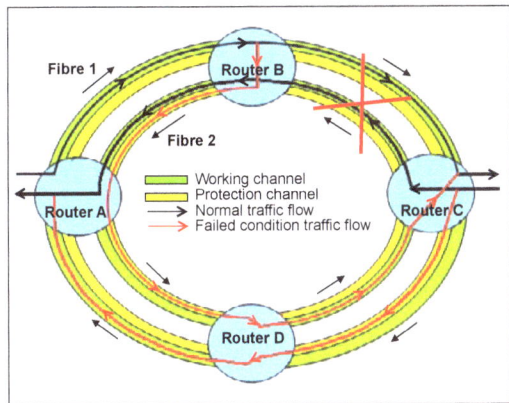

Fig. 6: Four-fibre bi-directional ring protection

an entire virtual ring is required for each path. In a BLSR, every link can carry both the working and protection traffic at the same time.

BLSR has two variants, namely, four-fibre BLSR and two-fibre BLSR. In a two-fibre BLSR, traffic is sent over both the fibres by utilising only half the capacity on each fibre and keeping rest half of the capacity reserved for protection.

In four-fibre BLSR, two fibres are used as working line and the other two are used for protection (Fig. 6). Each pair of fibres transports working and protection channels. This results in 1:1 protection, that is, 100 per cent redundancy. This improved protection is coupled with relatively high costs.

Ring and span switching mechanisms are used in the event of failure. In span switching, when the source or destination on a link fails, traffic gets routed onto the protection fibre between the two nodes on the same link. And when a fibre or cable-cut occurs, service is restored using the ring switching mechanism. Two-fibre BLSRs also benefit from the ring switching but cannot perform span switching like a four-fibre BLSR.

Let us take an example of network failure in four-fibre bidirectional ring. If a fault occurs between neighbouring elements B and C, network element C triggers protection switching and controls network element B by means of K1 and K2 bytes of the SDH system and traffic is re-routed as shown in Fig. 6.

Conclusion

Network failures are very crucial and have always been a concern of utmost importance. Such failures may result in heavy losses of traffic, leading to complete service outage. In order to safeguard networks from failures, automatic protection switching mechanisms are being widely deployed in long-haul networks as well as in inter-office networks due to their high-efficiency capabilities coupled with very-low (less than 50ms) restoration time. ●

The author, working with Bharat Sanchar Nigam Limited, holds Ph.D. degree in electronics engineering from Indian Institute of Technology (BHU), Varanasi. He has authored/co-authored more than 25 research papers in peer-reviewed national/ international journals. His current research interests include wired and wireless technologies for high-speed telecom services

How To Build An EMC Test Kit

All electronic devices have the potential to emit electromagnetic fields. In order to bring your product to market, you need to ensure it clears a set of EMC tests. Let us find out some facts about these tests and how to build a test kit for it

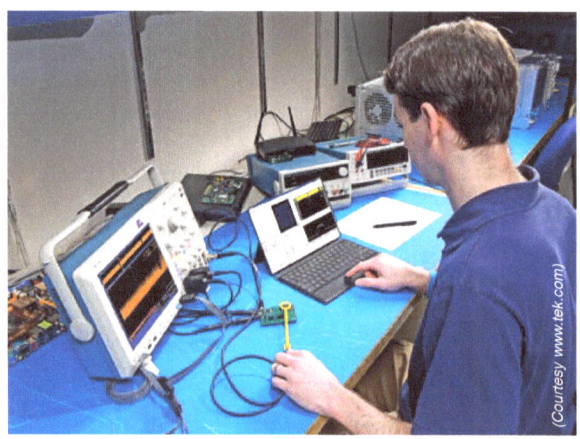

(Courtesy www.tek.com)

ANAGHA P.

Electromagnetic compatibility (EMC) testing, also referred to as EMC immunity testing, has become an important part of product design all over the world. Electronic equipment should meet certain mandatory EMC immunity criteria to be eligible for use in most countries. This ensures that the equipment will operate well when working in its intended application. Satisfying EMI/EMC safety standards, according to correct regulations, at design stage of the product can help pass the tests in one shot.

EMC testing

Let's start with a quick primer on EMC testing before moving on to selecting equipment for it. There are two types of emissions that are tested: conducted emissions (CE) and radiated emissions (RE).

CE test determines the radiations propagated along connectors and cables, causing noise in the system. A designer needs to check the susceptibility by choosing proper filters in conduction path. Emissions of all equipment need to be arrested, ensuring there are no disturbances to other equipment running on the same connectors and cables.

RE test is conducted to determine the disturbance caused to the device under test (DUT) in the presence of any radiation source. These can be arrested at the design stage. Radiated susceptibility testing is done to safeguard the device's immunity to strong radio transmitters in its vicinity. Product immunity testing is conducted to ensure human-machine safety.

How standards decide your equipment

Different types of equipment have different needs depending on factors like (a) operating environment (military equipment are used in harsh environments compared to consumer electronics), (b) reliability (medical devices that monitor vital signs should be more reliable than a personal computer), and (c) safety (if the operator has to be near the equipment, the radiation levels should be low). On the basis of these, the designer has to consider which specific industrial or military (MIL) safety standards should be fulfilled. However, simply knowing that you need to test for a specific industrial standard is not good enough.

Manufacturers of electronic equipment, unfortunately, have a big issue to face: These standards do not remain the same. They evolve to keep up with the pace and a requirement of modern technology on a continual basis. Some of these changes might be relatively minor, while others are withdrawn or completely re-written and superseded. It is important to keep up with all these changes and timelines for making sure a product reaches the market.

Why this trouble? The goal of the changing standards is to make the equipment become least intrusive for other devices, and more immune to hostile electromagnetic environment.

One issue is the increase in number of electronic devices that use advanced semiconductors. Electronic products are tending towards becoming more non-linear. The direct effect of this trend is seen as increased corruption of AC mains, which is a shared public utility.

Another problem faced is when previously discrete technologies converge in a single product. One example is the case of multimedia equipment for which a previously separate CISPR 32 standard was introduced. In order to maintain safety and best performance, the standards should be rewritten or replaced accordingly.

Additionally, with the evolution of technology for radio receivers (Bluetooth and its variants, Wi-Fi and its variants, cellular radio, etc) or electronic transmitters (FM transmitters, radars, etc), EMC standards should also change accordingly to cope up with the changing levels of EMI.

Is it EMI or EMC?

Electromagnetic interference (EMI) is the effect of electromagnetic inductions or electromagnetic radiations emitted by an external source on an electrical/electronic circuit.

Electromagnetic compatibility (EMC) deals with how well an electrical/electronic circuit works in an EMI environment—whether it resists generation of electromagnetic disturbances that may affect other products and living beings in the locality.

SWITCHING THE SMART WAY

Your Partner for Sensor Technology

Leading name in Reed Switch & Sensor Technology with more than 30 years of experience.

Superior quality Reed products, even for price-sensitive markets. Excellent support from concept stage through to series production.

Product overview

> Reed Switches

> SMD Reed Switches

> Reed Sensors

> Level Sensors

> Customized Sensors and PCB-Assemblies

For more information please visit pic-gmbh.com

Sensor Technology

Keeping up with changing standards

Designers, test labs and test equipment manufacturers should be very well aware of, and up-to-date, with the changing technology and standards.

Test equipment users must be familiar with the equipment provided by the vendors and their limitations. It is best to buy an instrument that offers programmability for test parameters. This helps you adjust the testing process to minor revisions of standards. Programmability also serves some customised, non-standard requirements that the customers would have adopted to make their products more reliable in a typical operating environment.

Test facility vendors may have to strengthen their equipment as per standards for the safety requirements. Most of the required upgrades would be with respect to sensitivity of the antennae, range of antennae, software updates of the test equipment and periodical calibrations.

"Specifically for the changing compliance standards, we proactively reach out to the designated authorities to know about the likely future compliance changes," says Raghu Rao, manager of application engineering team of Tektronix India. "This helps in being ahead in the market with future ready solutions."

Building a basic EMC troubleshooting kit

Troubleshooting is the process of isolating a problem and applying appropriate fixes. A test engineer diagnoses an EMI/EMC problem the same way a doctor diagnoses the medical condition of a patient. The diagnosis involves several stages: looking at clues, examining the equipment, gathering additional information (usually through tests) and finding a suitable solution.

An EMC troubleshooting kit is of great help to both independent and in-house EMC consultants. The essential tools for EMC troubleshooting depend on the kind of industry compliance the customer is looking for. But here is a brief guide that would help you

Whom to approach for testing in design stage

At the design stage itself, the user/designer has to estimate the size, power consumption, equipment required for testing and the range for which radiations and conductive emissions need to be tested. Vendors are selected depending on these parameters. For example, if a switch mode power supply (SMPS) or a radio system were to be tested, a small chamber with a minimum of Standardisation Testing and Quality Certification (STQC) Directorate approval would be sufficient. Volume of the chamber matters as the DUT should fit inside the closed chamber.

For preliminary tests, industries can develop their own test setups without any approval. This is done only for confirming the emission levels before taking an appointment with the certification body. Underwriters Laboratories (UL) facilitate testing and certification of standards for the industrial grade. But when it comes to testing of defence applications, one has to approach defence organisations like RCI in Hyderabad, who can test and certify for EMI/EMC and EMP (electromagnetic pulse).

Latest Standards for India

Product	Standard
Electronic games (video)	IS 616:2010
Laptops/Notebooks/Tablets	IS 13252:2010
Plasma /LCD /LED TVs of screen size 81cm and above	IS 616:2010
Optical disc players with built-in amplifiers (input power 200W and above)	IS 616:2010
Microwave ovens	IS 302-2-25:1994
Video monitors of screen size 81cm and above	IS 13252:2010
Printers, Plotters	IS 13252:2003
Scanners	IS 13252:2010
Wireless keyboards	IS 13252:2010
Telephone answering machines	IS 13252:2010
Amplifiers (input power 2000W and above)	IS 616:2010
Electronic musical systems (input power 200W and above)	IS 616:2010
Electronic clocks with mains power	IS 302-2:26: 1994
Set-top boxes	IS 13252:2010
Automatic data processing machines	IS 13252:2010

As per Electronics and Information Technology Goods (Requirements for Compulsory Registration) Order 2012, issued by Department of Electronics and IT (DeitY), India

assemble the most basic equipments required for basic troubleshooting, depending on your requirements, at minimum possible cost. The kit can be used for limited pre-compliance testing and assessing radiated emissions.

Spectrum analyser. The heart of an EMC troubleshooting kit is a spectrum analyser. Look for an instrument that is not bulky, preferably a handheld analyser, as it would be easier to carry around. It should have good external command support, fast triggering speed and a capability to buffer many triggered measurements (gated sweep). More useful, if you can afford, is real-time spectrum analyser. Conventional EMC testing involves analysing events

that are repetitive. The need for today's equipment is to be able to catch sporadic and rarely occurring events.

Probes. These include near-field probes and current probes. Probes can be constructed in lab with a little expertise in the field. H-field and E-field probes are near-field RF probes, constructed from semi-rigid coaxial cables, that are used to identify the source of electromagnetic interference emissions and potential radiation frequencies around circuits, cables and enclosures.

H-field probes are made basically from a conductive loop, and they detect magnetic fields produced by clock signals, control signals, serial data streams and switchers. A voltage is produced in the loop proportional to the magnetic field perpendicular to it. With larger size of loops you get higher sensitivity, but lower resolution. E-field probes, which make direct contact with the circuit, are used to find emissions on individual pins or PCB traces.

Poorly-bonded cables are said to be the major cause of radiated emission failures. A current probe, one of the most used accessories in troubleshooting, lets you measure the common-mode current flowing in wires or cables and identify a bad termination, which causes current leakage. Probes let you

Criteria for selecting EMC test instruments

1. The kind of task the user wants to perform. Requirements vary depending on the industry or application.
2. Working environment. If the equipment is for use in field, it needs to be heavy-duty, waterproof, rugged and portable. A benchtop instrument can perform delicate, precise measurements indoors.
3. Comprehensive coverage of test standard requirements so that the user can perform test to comply several standards using a single instrument.
4. Programmability and user friendliness of the test equipment.
5. Upgradability of the test equipment, in accordance with the changing standards or latest releases of the instrument, and local technical support available.

predict whether the given object will pass the emission test or not. For troubleshooting, high accuracy is not necessary, so you may go for low-cost probes or make your own current probes.

Antenna. A simple TV antenna positioned one or two metres away from the DUT would pick up harmonic signals. Connect it to the spectrum analyser to assess radiated emission. Changes below 2dB could be considered as measurement or setup error. Fixing the antenna in place using tape will reduce additional variables in measurement.

Preamplifier. A broadband pre-amplifier is used when the signal measured by the probes (particularly the smaller H-field probes) needs to be boosted in order to observe changes in emission levels.

ESD simulator. Electrostatic discharge (ESD) simulator or ESD gun generates ESD pulses that help verify the immunity of a particular device to static electricity discharges. While an actual ESD simulator can be expensive, you can create a rather simple ESD generator using a zipper storage bag with several coins inside. Shake this bag near the circuit to get several volts of ESD pulses at rise times of about 100 picoseconds.

ESD detector. Issues like loss of data and unusual circuit reset can be due to ESD in the circuit. Commercial ESD detectors are available in the market for higher accuracy applications. But an AM broadcast radio tuned off-station can act as a good ESD detector. If it has an FM receiver, you can even tune in product harmonics from radiated emissions.

Radiated immunity testing. Nothing can replace the test lab equipment for accurate radiated immunity levels. But a variable-RF signal generator would allow the simulation of radiated immunity test to some degree and give you an idea of whether the object is immune or not. Once the susceptible region is identified, you can implement potential fixes. This can save time and money as opposed to performing troubleshooting at test labs.

Other contents. Some of the other things include digital multimeter, screwdriver kit, pencil soldering iron, solder, SMA (subminiature version A) connector wrench, flash light, magnifier, tweezers, wires of different lengths and sizes, 10dB and 20dB attenuators, aluminium foils, coaxial adaptors, insulation tape, copper tape, measuring tape, EMI gaskets, ferrite chokes, common-mode chokes, and a range of I/O cables, BNC coaxial cable, SMA coaxial cable, resistors, capacitors and inductors.

"At a compliance lab level, the list of test instruments and test infrastructure can be exhaustive," says Rajneesh Raveendran, the project manager of Tarang Test Lab, Wipro. But to get a product certified in accordance with the appropriate and latest test standards you always need to seek help of an EMC testing lab. "Semi/fully anechoic chamber, EMI receivers, antennae and antenna masts, turntable, signal generators, RF amplifiers, line impedance stabilisation networks, coupling/decoupling networks, RF current probes, various immunity test generators, etc are provided to cater to the various radiated and conducted emissions/immunity," adds Rajneesh.

Three design elements that influence testing

The size of the DUT decides the measurement method and test duration for radiated emission and radiated immunity testing. As size of the object increases, the mechanical structure needs a good number of additional screws and holes to fix it. The additional slots and angles that come up to support the structure act as source of radiation, increasing the challenge faced by the engineer during compliance testing.

The complexity of its antenna pattern also increases with the size of the test object. "The antenna pattern complexity affects the number of orientations necessary to determine minimum immunity and maximum emission," says Vishal Gupta, application expert at Agilent Technologies India. This also has implications for EMC standards.

If your system's power consumption increases, it might result in the system requiring an active cooling system. This, in turn, demands for ventilation, creating openings that cause the radiation levels to increase, making it difficult to identify the source of radiation.

Similarly, with more number of cables, the tests will have to be done on each cable. Hence, for larger devices, more exposure is needed to measure gauge susceptibility and emissions. Moreover, the turntable or test chamber needs to be large to accommodate bigger equipment.

Very-near-field technique of EMC testing

The terms far-field, near-field and very-near-field describe the fields around an antenna. Near-field is less than one wavelength (λ) from the antenna and far field begins at a distance of 2λ and beyond. With the right implementation, any antenna can be successfully measured on either a near-field or far-field range. Ideally, far-field ranges are a better choice for lower frequency antennae and where simple pattern-cut measurements are required, and near-field ranges are better for higher frequency antennae and where complete pattern and polarisation measurements are required. The near-field is generally divided into two areas, the reactive and the radiative. The reactive region is what we call the very-near-field.

http://electronicasia.com

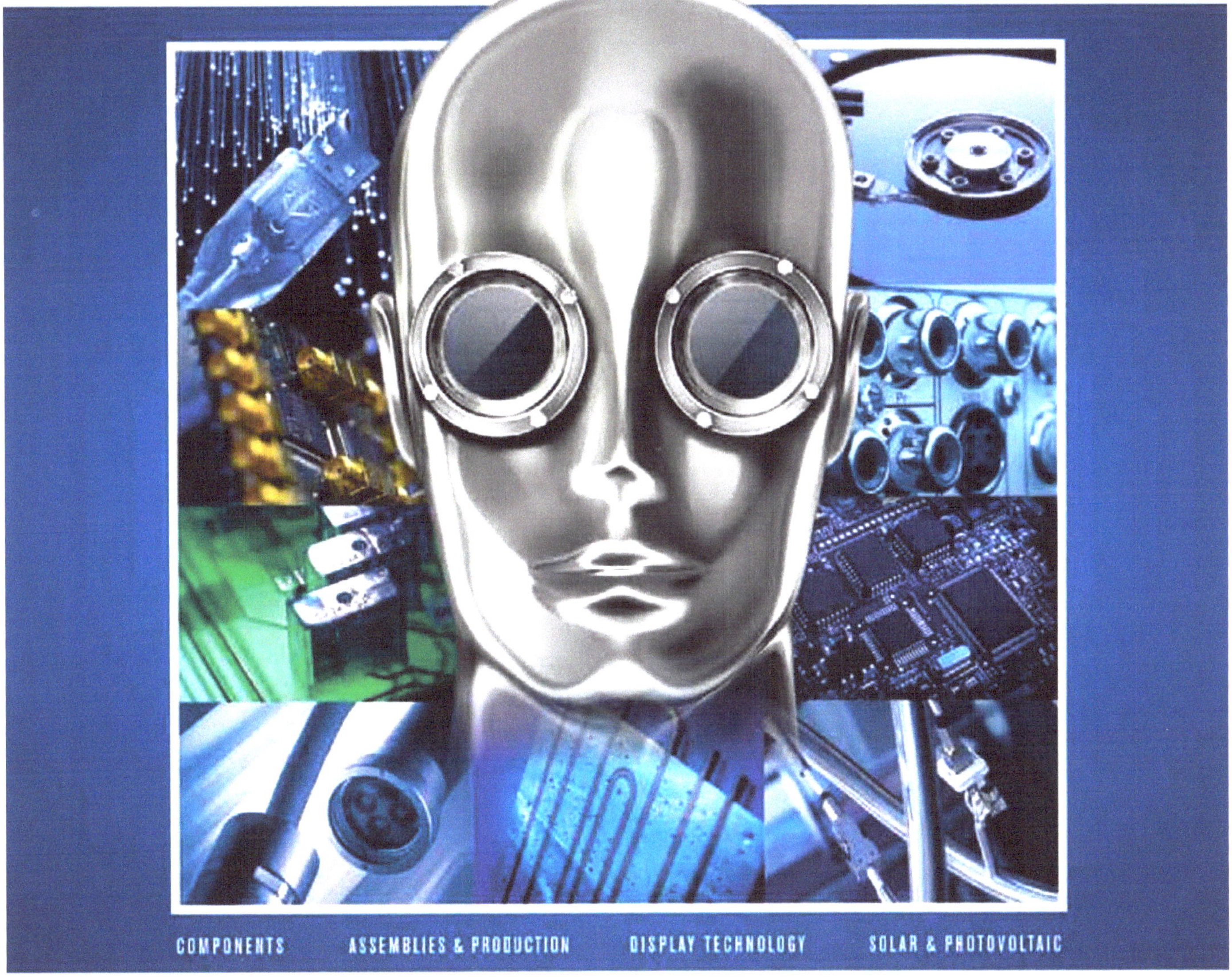

COMPONENTS   ASSEMBLIES & PRODUCTION   DISPLAY TECHNOLOGY   SOLAR & PHOTOVOLTAIC

electronicAsia 2014 ∞ufi

13 - 16 October 2014 • Hong Kong Convention and Exhibition Centre

In conjunction with HKTDC Hong Kong Electronics Fair (Autumn Edition)

Real Business through Global Marketplace

- Near 600 exhibitors from 12 countries and regions
- Highlights: World of Solar, World of Display Technology and Key Components for Smart Devices
- Concurrent with "HKTDC Hong Kong Electronics Fair (Autumn Edition)" – the world's biggest electronics event

Organisers :
HKTDC – HONG KONG TRADE DEVELOPMENT COUNCIL
Tel (852) 2584 4333 Fax (852) 2824 0026 Email exhibitions@hktdc.org

MMI ASIA PTE LTD
Tel (852) 2511 5199 Fax (852) 2511 5099 Email mmi_hk@mmiasia.com.sg

Reserve Your FREE Admission Badge!

For trade visitors, you can reserve your FREE admission badge through the following channels

Mobile Info Site – visit hktdc.com/wap/ea/T119.

Smartphone Info Site – download HKTDC Mobile at iPhone App Store, BlackBerry App World or Google Play, or

Website – http://electronicasia.com/ex/05

Quick, repeatable, real-time data. According to Erkan Ickam of EM-SCAN, a leading developer of fast magnetic very-near-field measurement applications, "Traditional EMC far-field techniques measure the emissions levels; they are not useful for small, printed circuit board assembly (PCBA) level diagnosis as they can't point at the source of emissions on the PCB." The very-near-field technique is fast and repeatable; interaction with DUT is unavoidable in this area. "It can measure the entire PCB continuously to make sure intermittent events are captured even when they occur in areas of the board that are not expected to radiate," he adds.

In other words, very-near-field technique is a good method to quickly identify localised hot spots at a PCBA level. It lets you identify frequency content of emissions and make spatial maps of the currents at each frequency as they occur on DUT. These spatial maps give information to test engineers about where the ultimate source of EMC emissions are, at the PCB level. These very-near-field spatial maps cannot be directly compared to any existing EMC standards like IEC and ANSI; they, however, provide insight into troubleshooting or root-cause analysis for any standard.

Complementary to all standards. Very-near-field testing is a pre-compliance tool and there are no standards for EMC pre-compliance. An example is the EMxpert by EMSCAN. Regardless of the standard, when current EMC chamber measurements indicate a failure, the test results do not generally indicate what the cause of the problem is. This is where a quick measurement utilising very-near-field is extremely valuable. It will indicate the source of emissions, coupling paths, resonant points, shielding issues as well as many other potential problems. This can help the engineers track down the problem quicker and make proper changes to the design.

Indoor testing. Another advantage cited for near-field measurement techniques is that testing can be accomplished indoors, eliminating problems due to weather, electromagnetic inter-ference, security concerns, etc.

Drawbacks. Each hot-spot seen in very-near-field probing is an independent vector by itself. Unless the magnitude and polarity of these independent vectors are quantified, convergence to real-lab measurements cannot be assured. Another issue is that the DUT must be smaller than the scanner for pre-compliance. The device should be disassembled for testing.

Testing at the component level

P. Chow Reddy suggests that, to test at component level, make a simple loop probe with a coaxial cable that could be connected to the spectrum analyser. One end of the cable can be stripped and centre wire extended to the braid of the shielded wire. This forms a complete loop with respect to the shield. This arrangement picks up radiations in the loop and transfers to spectrum analyser and hence gives a rough picture of the radiation level.

To specifically identify the radiating component, strip a thin-shielded wire with braid, make the centre wire into a small loop and solder it on to the braid. Then drop silicon sealant on the loop. This forms a mini probe. With one end connected to a spectrum analyser, this probe can be brought close to the specific components to analyse the pickup radiations.

Antennae tuned to particular range of frequencies and resonances are readily available in the market. These can be used to get a very brief picture of the radiations in open labs. ●

The author is a technical correspondent at EFY

Latest EMC test equipments

2171B Boresight antenna tower for ANSI standard C63.4 by ETS-Lindgren. Released in the first quarter of 2014, model 2171B antenna tower is designed to meet ANSI standard C63.4 requirements for measurements above 1GHz. The patented Boresight system keeps the antenna aimed at the DUT during the antenna's ascent or descent along the antenna mast.

Red Edge series for electromagnetic pulse protection by ETS-Lindgren. This product line, announced in the last quarter of 2013, helps negate high-altitude electromagnetic pulse (HEMP) and intentional electromagnetic interference (IEMI) threats.

Double-ridged waveguide horn and BiConiLog antennae 3116C-PA and 3117-PA with matched preamplifiers by ETS-Lindgren. The antennae are available with matched preamplifier since the third quarter of 2013, both calibrated as a single unit. All the possible mismatches between the antenna and the preamplifier would be taken into account in the antenna factor (AF) which is provided with each unit.

CEMS100 test platform for EMS (electromagnetic susceptibility) testing for IEC/EN 61000-4-3 by Rohde & Schwarz. This is a flexible and reliable off-the-shelf solution for EMS measurements in line with IEC/EN 61000-4-3 standard, announced during the first quarter of 2014. The platform includes all the components needed for EMS and EMI measurements, covers common frequency ranges and field strengths required for pre-compliance tests and certification.

ESRP EMI test receiver by Rohde & Schwarz. The device has EMI test receiver and signal/spectrum analyser combined in one box, and offers optional pre-selection and preamplifier. It was released during the first quarter of 2014.

Line impedance stabilisation networks (LISNs)/artificial mains networks (AMNs) series by TDK RF. This new line of LISNs/AMNs, announced in the first quarter of 2014, can be used for various conducted EMI testing applications. These are offered separately as well as integrated into compliant and pre-compliant solutions for automated EMI measurements.

MDO3000 series of mixed-domain oscilloscopes (with integrated spectrum analyser) by Tektronix. This customisable and upgradable 6-in-1 integrated oscilloscope includes integrated spectrum analyser, logic analyser, protocol analyser, arbitrary function generator and digital voltmeter/counter. It was released in the first quarter of 2014.

RSA5000 spectrum analyser by Tektronix. A mid-range spectrum analyser released during the last quarter of 2013. It has features like DPX Live RF spectrum display, seamless data capture, and automatic pulse measurement and detection.

MAJOR CONTRIBUTORS TO THIS STORY

- ➤ **Erkan T. Ickam,** director of marketing, EMSCAN
- ➤ **Kenneth Wyatt,** senior EMC engineer, Wyatt Technical Services, LLC
- ➤ **P. Chow Reddy,** manager-R&D (Power Division), ICOMM Tele
- ➤ **Raghu Rao,** manager of application engineering team, India, Tektronix
- ➤ **Rajneesh Raveendran,** project manager, Tarang Lab, Wipro
- ➤ **Vishal Gupta,** application expert, Agilent Technologies India

In electro-chemical solutions, we reign

In the formulation, manufacture and supply of conformal coatings, thermal pastes, encapsulants, cleaners and lubricants, we have the solution. Through collaboration and research, we're developing new, environmentally friendly products for many of the world's best known industrial and domestic manufacturers – always to ISO standards.

Combine this unique ability to offer the complete solution with our global presence and you have a more reliable supply chain and a security of scale that ensures you receive an exemplary service.

Polyurethane & epoxy resins

Thermal management solutions

Contact lubricants

Water and solvent based cleaning

Conformal coatings

Maintenance & service aids

Conformal Coatings

Enhanced protection for enhanced PCB performance

ELECTROLUBE
THE SOLUTIONS PEOPLE

Conformal coatings are designed to protect printed circuit boards and related equipment from their environment. They conform to the contours of the board allowing for excellent protection and coverage, ultimately extending the working life of the PCB. They protect form moisture, salt spray, chemicals and temperature extremes in order to prevent corrosion, mould growth and electrical failure. Electrolube has the solution.

- UL, MIL and IPC-CC-830 approved
- Solvent removable and solvent resistance coating
- Acrylic, Silicone, Polyurethane and Hybrid Materials
- UV cure and water-based options available
- UV trace to aid inspection
- Thinners and masking products

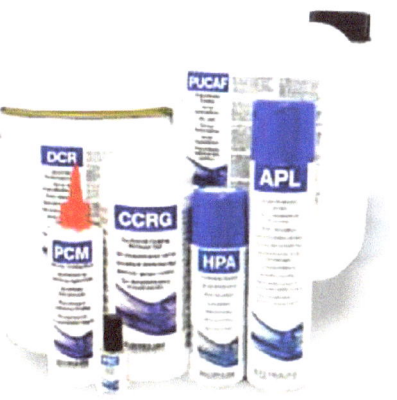

Isn't it time you discovered how Electrolube can serve you? Simply call, or visit our website.

What's New in Displays for
Embedded Systems

Various challenges, such as physical size and power consumption, have become more of a concern for embedded devices. This article presents some high-level solutions that are available to tackle the challenge

PANKAJ V.

We have all seen the transition from CCFT- to LED-backlit LCD modules, and the many benefits it provides for design engineers working on incorporating display technology into their devices. Modern displays, such as TFT and OLED, are evolving with promising designs for tomorrow, including the integration of new high-speed interfaces to handle the emergence of very-high-resolution imaging needs and the possible elimination of backlight in E-Ink displays. Overall, displays for embedded systems have evolved well enough to help design efficient and better products, often directly affecting the duration of the design cycle itself. Let's see how.

Ready-to-deploy solutions

The choice of which embedded display to use could be a complex decision to figure out, especially for those who are new to integrating an embedded display into their design. Vendors have announced new all-in-one display products that cater to the requirements of a design engineer developing from industrial to consumer applications, with the main aim being to make the design process faster. These displays are engineered to integrate anything from power supplies to integrated processors. They include all the necessary components and integrate various functions like switch, UART, USB, SPI, I²C and PWM, along with SD card storage, to allow for large storage capabilities.

Ravi Pagar, regional director–South Asia, element14 India, says, "The all-in-one offerings reduce time-to-market and development costs. Engineers can quickly integrate multiple functionalities into their next design and also upgrade their designs, quite simply. These ready-to-deploy solutions can be adopted for products in various areas like industrial, consumer electronics and telecommunications."

VIA launches new series of rugged LED backlit displays for embedded applications

Upgrade for mono STN LCD modules. If you do not have much experience on TFT module development, and you want to upgrade your mono LCD display to a TFT, there is no need to worry about microcontroller unit (MCU) limitations. TFTs with integrated processors are available for upgrading mono displays to colour TFT displays, saving R&D engineers time and accelerate the product development procedures.

These latest 'clever' TFTs do not require any modification to the exist-ing firmware and circuit on your PCB. For instance, Winstar's WF35M allows the designer to implement colour TFT without having to redesign the TFT firmware and circuit. You can easily switch to TFT with a simple procedure, avoiding too many I/O, without using an expensive high-performance CPU or MCU. These TFTs eliminate the need for using FPC at the client's end, thereby adding to the savings for your bill of material (BOM) costs for final products.

Onboard video processing for easier integration. Systems having a combination of video processing functionality built directly into the screens have made it possible for easier integration with a number of applications. These display modules, like 4D System's 4DLCD-FT843 and FTDI's VM800B, are highly advanced intelligent SPI displays, based on integrated video engine and amplifier-enabled audio, providing low-cost display solutions for your embedded designs.

The integrated engine offloads the host processor and provides a variety of graphics features. Besides, various interfacing options allow different hosts to be connected directly to the displays, providing a powerful set of audio and graphics features to the host using the on-board audio/video engine. These modules are best suited for requirements of various HMI applications, including industrial control terminals, intelligent instruments, data acquisition and analysis, medical products and network terminals.

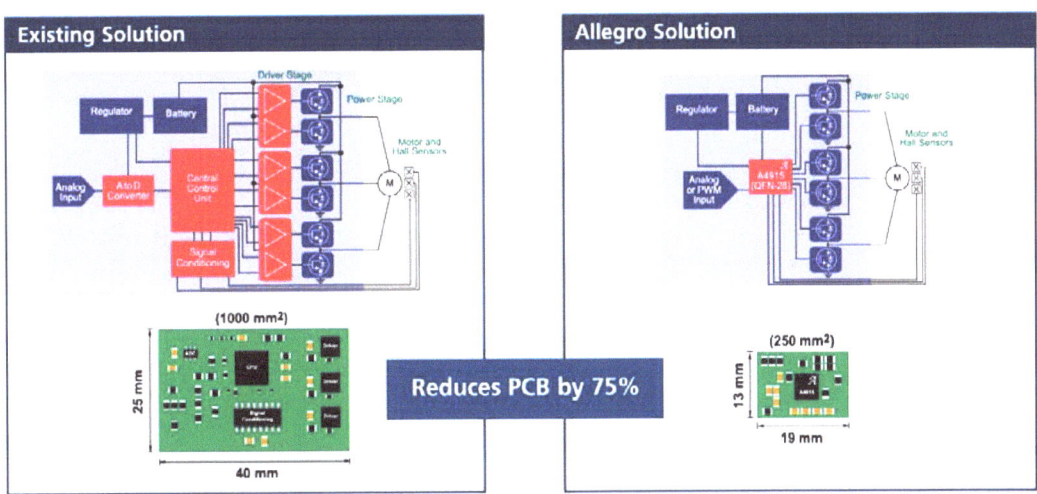

Latest Display Modules

Module	Manufacturer	Resolution	Dimensions (WxHxD mm)	Brightness (cd/m²)	Interface	Contrast ratio	Viewing angle (U/D/L/R)	Type
DET035QVNTCMI-1W	Densitron	320x240	76.9x63.9x4.85	500	I2C	800:1	80/80/80/80	Multi-touch TFT module
WF62A	Winstar	640x320	170.32x88.3x5.3	1000	MCU	800:1	80/80/80/80	TFT module
AA070ME11	Mitsubishi Electric	800x480	118.5x84.7x3.9	1500	LVDS	600:1	60/80/ 80/80	TFT-LCD module
AA050MG04	Mitsubishi Electric	800x480	118.5x84.7x3.9	1000	CMOS	1000:1	85/85/ 85/85	TFT-LCD module
TX17D200VM0BAA	KOE	800x480	154.8x92.7x10.2	700	LVDS	600:1	85/85/85/85	TFT module
AMS465GS45	Samsung	720x1280		300	LVDS	3400:1	80/80/80/80	OLED module
H499TLB01 V0	AUO	720x1280		250	MIPI	1000:1	80/80/80/80	OLED module
DLA320480AB035F	Densitron	320xRGBx480	52.9x83.5x1.22	180	MIPI-DBI 8/16/24 bit MPU/RGB, MIPI-DBI Serial			OLED module

TFTs with enhanced optics

Displays combining exceptional projected-capacitive-touch (PCT) technology with industrial-grade TFTs are able to offer higher transmissivity and unlimited controller-dependent multi-touch capability. Now we have a combination of high resolutions with extremely compact form factors up to depths as low as 4.6mm, enabling the design of even thinner devices or providing space for integrating more components.

Modules like those in Densitron's DET035QVN series have improved TFTs with ultra-wide-view (UWV) polariser solution, increased viewing angles in each direction with greatest colour consistency, which is well suited for use in high-end, graphics-rich applications, for both fixed and mobile devices. And they consume as low as 2.8W.

If we talk about mono TFTs, the latest monochrome models with TN-positive LCD applied in the active-matrix TFTs are able to provide brightness as high as 1000-1500 nits, and high contrasts as much as 800:1 with short response times. These models, like the Winstar's Mono WF57S, are superior alternatives to traditional monochrome LCD modules.

Thin-glass substrate EPD. Based on new thin-glass TFT technology, the latest E-Ink or electronic paper display (EPD) modules are catering to the demands of portable consumer electronics products that are much lighter and thinner as compared to other standard displays. These weigh around 50 per cent lighter than an equivalent glass-based TFT and are 50 per cent thinner.

Reflective displays for lower-power applications

For any embedded application, primary concern is power. Most displays use a backlight that is projected through the display. This backlight consumes most of the power. Therefore eliminating the need for any backlight in displays will surely increase the battery life of the products.

In contrast to the backlight technology, we have power-efficient reflective-technology-based E-Ink displays. These do not require any kind of backlight and take ambient light from the environment, which is reflected from the surface of the display, thereby saving a huge amount of power.

E-Ink: What adds to the power efficiency

Another advantage of E-Ink displays, which adds to the power efficiency of their products, is the bistability of an EPD. An image on the E-Ink display screen is retained even if all the power sources are removed. Thus, these display modules could be used to make the most power-efficient displays that will consume power only when something is changing and not when it is idle.

No idle refreshing is required in E-Ink displays, while any traditional LCD display needs to be refreshed around 30 times per second, regardless of whether or not anything new is being displayed.

In addition to being thin and light, these modules deliver the same energy efficiency and daylight readability of all E-Ink displays. Besides, the energy efficiency further reduces weight of the end products by requiring a much smaller battery.

Flexible active-matrix displays for wearable electronics. MOBIUS is another E-Ink technology that is mainly used by smartwatch start-ups like Sonostar and Transmart for designing flexible wearable display solutions. MOBIUS-based modules are smaller version of displays based on flexible-TFT technology. Lightweight and rugged properties make them conformable, so the end product has a better fit for the consumer.

With the inherent power management capability, these EPD modules are ideal for lightweight portable products that require a large display surface. Also, if you only want to display some numbers, and those numbers change very infrequently, no point in refreshing the display every time to burn power.

OLED technology for high-end mobile devices

OLED technology has been around for quite some time, but now it is evolving with more advanced features and physics, promising to be the technology for the future designs. OLED displays are improving at a very impressive rate of about 20 per cent or more year-on-year in brightness, colour management, colour accuracy, resolution, PPI and power efficiency.

OLED technology has many advantages over the other display technologies in terms of performance as well as power. Avinash Babu, senior architect, Mistral Solutions, explains, "People are slowly moving the display

Agilent Technologies

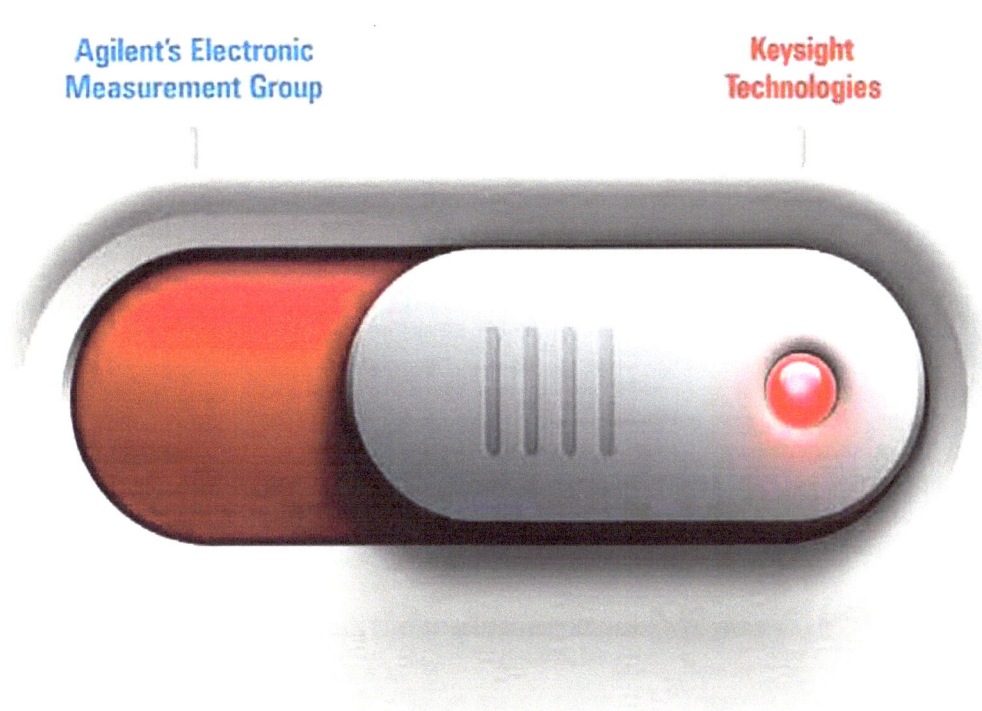

Agilent's Electronic
Measurement Group

Keysight
Technologies

Agilent's Electronic Measurement Group,
including its 9,500 employees and 12,000
products, is becoming **Keysight Technologies.**

Learn more at **www.keysight.com**

Toll-free: +1-800-112-929

   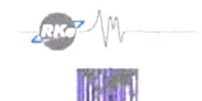 

technologies to OLED-based technology. An OLED-based display generally helps the designers in terms of having a better contrast, it is lightweight and it consumes less power. If you see the viewing angle as well, it is better in OLED as compared to TFT."

Flexible and curved OLED modules. Special OLED solutions are available for design engineers, making it possible to have curved and flexible displays for the futuristic designs. It is one of the major and important innovations in the display world, helping to design better and efficient displays. The popular Samsung YOUM line of displays targets designers of smartphones and wearables.

Curved surface significantly reduces, and sometimes even eliminates, reflections from ambient light sources. This not only improves screen readability and image quality but also allows the displays to run at lower brightness, thereby increasing their power efficiency and battery-running time. These OLEDs, being incredibly thin at just a fraction of a millimetre, along with the curved and flexible surfaces, are a suitable choice for mobile displays and the trending wearable displays.

Even higher resolution and brightness. As if resolution on current Samsung Galaxy S5 wasn't enough, many vendors have introduced even-higher-resolution AMOLED panels and modules. One example is AUO's WQHD AMOLED panel with 2560×1440 resolution at 513PPI. But vendors like Sharp, Samsung and SDC have gone further and announced resolutions of 560, 664 and even 860PPI. Apart from high resolution, ultra-slim designs up to 0.57mm have been realised and special drivers are being integrated to increase their touch sensitivity.

Using new structure and new OLEDs, manufacturers have come up with high-brightness PMOLED modules, which mainly target the increasing demand for wearable devices. These new OLED modules offer unparalleled clarity, even in direct sunlight, featuring a luminance of 1000 nits, which is nearly ten times of that

available with average OLED modules.

The only challenge for OLED technology is the manufacturing cost and yield. Though OLEDs have many benefits associated with them, manufacturing of OLEDs is still lacking behind in terms of productivity. Nate Srinath, founder-director, Inxee says, "Cost of manufacturing OLEDs is much higher than LCDs, and also people still don't appreciate their significance. The major challenges for OLEDs are reducing production costs and increasing production capacities."

Light capturing cavities in numeric displays

Demand for compact and efficient products has resulted in more compact and space-saving designs in 7-segment numeric displays, like those found in Lumex LDP-2R2608RD-50 series. These latest LED-based offerings use the brightest chip technology and light-capturing cavities, which allows them to offer maximum brightness and legibility without sacrificing any colour intensity or uniformity.

These single-package designs with multiple digits and icons can display any type of word or icon due to the display overlay. If you are looking for displays for board or panel indication, telecom switches and central station equipment, control panels with high-brightness lighting, these compact

solutions might be the answer.

Developing displays for embedded systems

There are various standard low-cost display development kits available to aid your designs. These kits not only have the required hardware like display module, cables, controllers and power supply but also the reference manuals and codes (if required) to program your modules. You can also have live online support for working with the kits. Some vendors provide an exclusive GUI web-portal development site license along with their kits to help design engineers with an advanced approach to web-based development.

Using the latest hot pluggable USB-based OLED development kits, you can drive OLED displays from the USB port of a PC. These are extremely easy to use yet powerful demonstration tools that do not require any extra cable or power supply to run, allowing displays to be up and running in minutes. A kit will provide you with a USB controller card, mini USB cable, an interchangeable OLED display card and a CD with software application and drivers. Also available are the display expansion boards and capes for various development board environments. Besides, some vendors provide customised solutions with kit modules to quickly interface the dis-

---

Toshiba, a world leader in semiconductor technology, offers an extensive line of MCU for **MOTOR CONTROL** Application.

Product lineup is arranged according to the Motor Controlling Method and Application

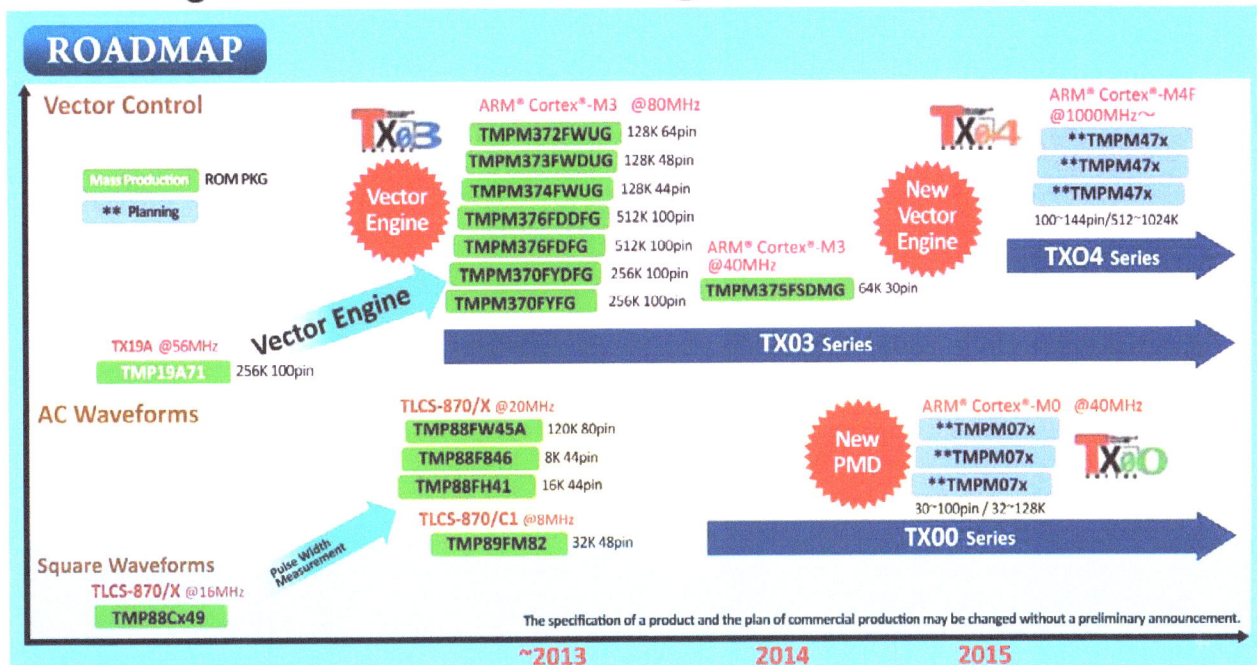

ROADMAP

Vector Control

ARM® Cortex®-M3 @80MHz
- TMPM372FWUG — 128K 64pin
- TMPM373FWDUG — 128K 48pin
- TMPM374FWUG — 128K 44pin
- TMPM376FDDFG — 512K 100pin
- TMPM376FDFG — 512K 100pin
- TMPM370FYDFG — 256K 100pin
- TMPM370FYFG — 256K 100pin

ARM® Cortex®-M3 @40MHz
- TMPM375FSDMG — 64K 30pin

ARM® Cortex®-M4F @1000MHz~
- **TMPM47x
- **TMPM47x
- **TMPM47x
100~144pin/512~1024K

- Mass Production — ROM PKG
- ** Planning

Vector Engine

New Vector Engine

TXO4 Series

TX03 Series

TX19A @56MHz
- TMP19A71 — 256K 100pin

AC Waveforms

TLCS-870/X @20MHz
- TMP88FW45A — 120K 80pin
- TMP88F846 — 8K 44pin
- TMP88FH41 — 16K 44pin

TLCS-870/C1 @8MHz
- TMP89FM82 — 32K 48pin

New PMD

ARM® Cortex®-M0 @40MHz
- **TMPM07x
- **TMPM07x
- **TMPM07x
30~100pin / 32~128K

TX00 Series

Square Waveforms
TLCS-870/X @16MHz
- TMP88Cx49

Pulse Width Measurement

The specification of a product and the plan of commercial production may be changed without a preliminary announcement.

~2013   2014   2015

STARTER KITS

Starter Kit for TMPM370

Mass Production

Contents — Evaluation Board for TMPM370
J-LINK for TOSHIBA (or i-jet Lite)

Manufactured by IAR systems AB

New Product

Starter Kit for TMPM370

Contents — Evaluation Board for TMPM375 loaded inverter circuit
i-jet Lite
DC Motor

The TMPM370FY is ideal for motion control in home appliances and industrial applications where it reduces the need for additional components while providing significant benefits over software-based vector control running on a microcontroller.

Starter Kits also available from NGX Technologies Pvt. Ltd. Email: sales@ngxtechnologies.com

---

TOSHIBA
Leading Innovation >>>

Semiconductor & Storage Products
Contact at: Arun Kumar : +91-9810121489, Email: arun.kumar@toshiba-india.com
Website: http://www.semicon.toshiba.co.jp/eng/index.html

play module to your Raspberry Pi or Arduino-compatible board.

If you are working with E-Ink displays, there are easy-to-use PC USB powered kits based on open source software and tools. The latest EPD kits include pre-loaded firmware to take data via USB that are later sent over to the extension boards.

Future: LED displays with high resolution

When we talk of LED displays, these are not true LEDs. These are LCD displays with LED backlights. Nate Srinath says, "Till date we haven't perfected an LED display with very high resolution."

We can increase features in TFT-LCD technology and make power-efficient designs, but we have to look for advancements in greener technologies like LED and OLED for the future. Previously it was possible to make only small OLED displays, but over time manufacturers have started making the process more efficient. They are now able to make larger

MAJOR CONTRIBUTORS TO THIS REPORT

Avinash Babu
senior architect,
Mistral Solutions

Farheen Ali
director, Oriole
Electronics Pvt Ltd

Nate Srinath
founder-director,
Inxee

Ravi Pagar
regional director-
South Asia,
Element14

Sunil Khetwani
branch manager-
Mumbai, Winstar
Display Co Ltd

displays, but still the cost-efficient yield of OLEDs has to be achieved.

Experts believe that true LED displays with high resolution can be more power-efficient with a high contrast and brightness coming as natural features. However, can we actually have high resolutions with true LED displays?

Nate Srinath, founder-director, Inxee says, "It is possible with manufacturing capability changes. Today we can't manufacture an LED lamp smaller than a particular dimension. So we have to improve upon that and make smaller and smaller LEDs to hit extremely minute dimensions and be able to control them individually with good spectral characteristics. This will eventually lead to displays that will be very rugged and extremely modular. You can build your own giant flat-screen TV using smaller modules. Even if you throw water or stones on them, or break a pixel, it will continue to operate. Maybe 4-5 years down the road, we will be able to see such displays." ●

The author is a technical journalist at EFY

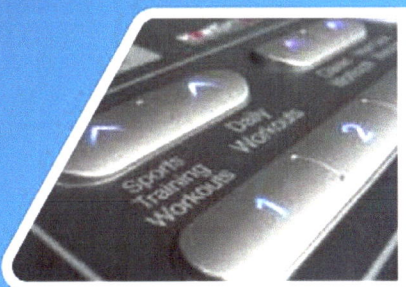  

The IoT is on Everybody's Mind

The Internet of Things (IoT) appears to be one of the hottest topics in the IT world today and seems to be significantly influencing the strategies and plans of companies in varied segments ranging from mobile and data centre to systems security. Here is a collation of some recent strategy-related announcements from industry leaders that shows how much companies are betting on the IoT. In fact, at major events including the International CES and Mobile World Congress, people seemed to be talking about little else

JANANI GOPALAKRISHNAN VIKRAM

Intel Inside. For a long time, ARM has been (and continues to be) inside many a successful IoT devices, but now Intel is gearing up for the race. In February this year, Intel revealed an extended portfolio of computing and communication assets, focussed on building the IoT. "The continued growth of the mobile ecosystem depends on solving tough computing challenges—unlocking data's potential while securely and reliably connecting billions of devices with leading edge computing and communications technologies," said Intel President Renee James at the Mobile World Congress.

"Today we are announcing leading communications products as well as new computing platforms. As a result, Intel is well positioned to shape the future of mobile computing and the Internet of Things." Amongst the products on show at the event was Intel's 2.13GHz Atom processor Z3480 for Android smartphones and tablets, a 22-nanometre Silvermont microarchitecture based 64-bit ready SoC with Intel's new Integrated Sensor Solution, which efficiently manages sensor data to help devices remain smart and contextually-aware even in a low-power state.

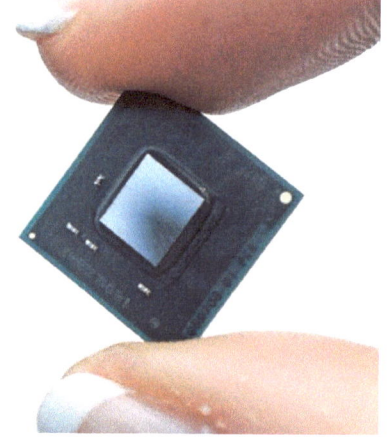

Intel Quark SoC X1000—the first product from the Intel Quark technology family of low-power, small-core products. Intel Quark technology will extend Intel architecture into rapidly growing areas—from the Internet of Things to wearable computing in the future (Courtesy: Intel)

Intel also spoke about its upcoming 64-bit processor Moorefield and stronger LTE-Advanced capabilities in upcoming products. Some weeks later, Intel also introduced the Intel Gateway Solutions for IoT, a family of platforms based on Intel Quark and Atom processors, along with McAfee and Wind River software, for companies to seamlessly weave industrial devices into an IoT-ready system of systems.

McAfee, securing the IoT. According to IDC, the installed base of the Internet of Things will be around 212 billion 'things' globally by the end of 2020. Put differently, that is, approximately 27 things per person! While the myriad devices ranging from smartphones and mobile devices to industrial sensors provide a rich user experience and enhanced effectiveness, they also pose serious threats to privacy and security.

McAfee hopes to secure all these devices smoothly without hassling the users, while enabling them to perform effectively. Their strategy comprises several dimensions, including assurance that devices are operating as intended by the manufacturer and have not been corrupted; lifecycle security across the device, network and data centre; support for industry standards and device interoperability; ability to solve IT/cloud services challenges in connecting legacy and new systems to new and future services; and technology to assure individual privacy. McAfee is working closely with Intel and Wind River in achieving this goal.

Cisco, enabling smartness. Speaking at their annual press event, Rob Lloyd, president of sales and development at Cisco mentioned that they see a $14.4 trillion business opportunity in the IoT, with initial markets being manufacturing, government, energy and healthcare. The company is working with several partners in various verticals to tap these opportunities.

In the energy space, they are working with utilities, hoping to deploy as many as 10 million smart meters supporting Internet Protocol (IP) by the end of the year as against current figures of 2 billion smart meters in operation using 135 varied utility protocols. They are

As the IoT picks up, researchers are figuring out ways to do away with the touch screen and use the body as the user interface (Courtesy: http:// thenetwork.cisco.com)

also working on migration strategies to smoothen deployment. On the tech front, Cisco has fit an IPv6 stack into a 40kB RAM. The company has also invested in Cohda Wireless, an 802.11p board vendor playing a key role in connecting cars. This and other Cisco-enabled products, such as Streetline, a parking management device, are examples of sensors that will create wondrous opportunities for Big Data players.

Microsoft, Internet of Your Things. Some time ago, Microsoft announced a Windows for IoT platform. But more exciting is the company's acquisition of Azure Intelligent Systems, a key move towards making the IoT more accessible. According to a press statement by a company official, Microsoft is thinking about the Internet of your things and not the Internet of every-

thing, meaning they wish to make the whole concept more user-friendly and accessible to users. How would Azure Intelligent Systems help? It is a solution for people to manage their entire IoT infrastructure from a single site. It is essentially a Cloud-based service designed to manage IoT devices, collect the data they produce and route it to useful tools, including their own Hadoop-based HD Insight, Office 365's Power BI, as well as custom software from Microsoft partners.

Intelligent Systems is a cross-platform service that augments existing infrastructure investments. It connects to devices over IP, and lets you connect to any device, anywhere and get data off that device. Even Linux devices! That is what makes the acquisition critical to Microsoft's IoT strategy. While Windows for IoT helps position Windows as a player in the IoT space, Intelligent Systems helps them span the whole market.

Oracle, small devices to Big Data. Oracle is also betting big on the IoT. Java-enabled devices and engineered Big Data systems are central to the company's plans. They are positioning embedded Java technology as a key enabler of the IoT, and develop Java-based solutions for small-to-large embedded devices, helping reduce costs, improve product quality and speed time-to-market.

While embedded devices are one end of the IoT spectrum, Big Data infrastructure is the other and Oracle has a strategy for that too. The company designs 'engineered systems' that are

Monetising the IoT

While companies are slowly gaining awareness about the IoT, many are still wary about how the infrastructure can actually be monetised. In an inspiring example, Transwestern and Metropolis Investment Holdings, well known in the property management space, renovated their business building systems with sensors and saw quantifiable results. They connected people, processes, data and things to a building-wide IP network.

By integrating everything into a backbone, they could monitor and manage 95,000 network-connected building points centrally, thereby enhancing building efficiency and bringing down operating costs. They could also respond to tenant requests in real time, improving customer experience. Also, by using business intelligence to study energy patterns, they could cut annual energy costs by 21 per cent in the first year.

Why the phone?

In many ways, the smartphone is becoming central to the IoT, with almost every device being connected to a mobile app. How safe is that? In response to a blog post about some IoT gadgets, a reader had replied worriedly that the mobile phone was the object he tended to lose most easily. Now, does that ring a bell for any of you?

pre-integrated to reduce the cost and complexity of IT infrastructures while increasing productivity and performance. ●

The author is a technically-qualified freelance writer, editor and hands-on mom based in Chennai

Read more about the Internet of Things at internetofthings.electronicsforu.com

When Machines Talk to Machines

During an interview on November 30, 2013, Jeff Bezos, CEO, Amazon.com took the correspondent into a secret room at the Amazon office and exposed an R&D project—Octocopter—drones that will fly delivery packets straight to the customers

Prime Air drone proposed by Amazon (Source: http://www.computerworld.com.au)

DEEPAK HALAN

Given that safety tests and Federal Aviation Administration approvals are required for flying octocopters to deliver packets, the technology could be available to customers by 2017. Shortening the delivery times is becoming a USP for online retailers. UPS, Dominos, eBay and Tesco are some other companies that are thinking on the same lines.

UPS is toying with the idea of using drones not only for delivering to customers but also for improvement in warehouse operations and bringing packages from aircraft to delivery trucks more effectively.

In time to come, your air-conditioner and washing machine will be able to 'communicate' to a smart meter installed in your home. The meter will automatically negotiate for the best rates with the energy providers. The smart meter will then remotely control your home appliances, automatically turning them on at the most favourable times to help conserve precious energy. This will bring down your electricity bills and improve communication between the energy providers and you.

In some years from now, you would be able to breathe fresh air on the roads. Smart electric cars will provide drivers with the latest traffic congestion and best-route information. You will be alerted before your vehicle breaks down. Also, you will be able to instantly communicate with emergency services. These are not scenes filmed in sci-fi movies, but trends that will be made possible by machine-to-machine (M2M) technology in the near future.

What is M2M

In simple words, M2M technology enables communication between smart devices connected via Internet. M2M is about enabling the flow of data between machines and machines and finally machines and human beings.

Irrespective of the type of machine or data, M2M usually involves a device (for example, a smart electricity meter) to record a parameter (such as electricity rate) which is relayed through a network (wireless such as Internet, wired such as telephone lines or hybrid) to an application (customised software program) that converts the recorded parameter into meaningful data (such as whether the washing machine should be switched on now or later when the energy rate is likely to be better). What varies within this basic framework are the various ways in which a machine is connected, the type of communication method used and the manner in which the information or data is utilised.

The four basic phases in an M2M application are collection of data, transmission of selected data through a communication network, assessment of the data and response to the available information.

Usually, an intelligent wireless data module is physically integrated with the monitored machine and programmed to understand the machine's protocol (the way it sends and receives

Several automobile firms have sensors that send information directly from the shop floor to the production plant's computers (Source: Kazuhiro Nogil/AFP/Getty Images)

data). In case the monitored machine is constituted as an intelligent master device, it could use the M2M device as a basic wireless modem. The M2M device would be fed data and then told to send it to the network. However, if the monitored machine is a mere bunch of switches and sensors, or is an intelligent slave device, the M2M device will then serve as the master device and regularly take inputs from the sensors and switches, or send data requests via the serial port.

Getting on to the cellular or satellite networks generally needs a gateway. A gateway receives the data from the wireless communication network and transforms it so that it can be sent to the network operation centre. Data security aspects, such as authentication and access control, are taken care of by the gateway and the application software.

Key M2M application areas

Today, M2M technologies have entered several walks of life. Energy, transport, real estate and agriculture are some major areas where this technology is fairly evolved. Benefits of M2M go beyond making life easier and smarter—they also include sustainable development of our planet.

Global warming is a serious environmental concern increasingly threatening our planet. The higher the CO_2 and other greenhouse gas (GHG) emissions, the greater is the global warming impact. Hence it is important that we cut down GHG emissions as much as possible, and M2M technologies help in achieving this goal too.

Energy. A smart meter is an electrical meter that records consumption of electric energy in intervals of an hour or less and transmits this data at least every day back to the utility central system for monitoring and billing purposes.

Hence smart meters enable two-way communication between the meter and the central system using M2M technology and form a part of advanced metering infrastructure. This leads to a smart grid which is a

A smart meter (Source: http://smartgridtech.wordpress.com)

modernised electrical grid that uses ICT to collect, interpret and act on information—such as demand pattern of consumers and supply pattern of the utility providers in an automated manner.

Time-of-day pricing, demand management, load balance and load optimisation help to improve the efficiency, reliability, economics and sustainability of the production and distribution of electricity.

M2M can also raise energy efficiency levels with regards to production and transmission, and hence further bring down CO_2 emissions by enabling usage of more renewable fuel resources.

It is estimated that M2M could save more than 2.0Gt of carbon dioxide emissions by 2020 in the energy sector alone by enabling the adoption of 'smart grid' technologies for both small and large-scale users, including smart meters and demand response systems.

Transport. In the transport sector, M2M applications range from the

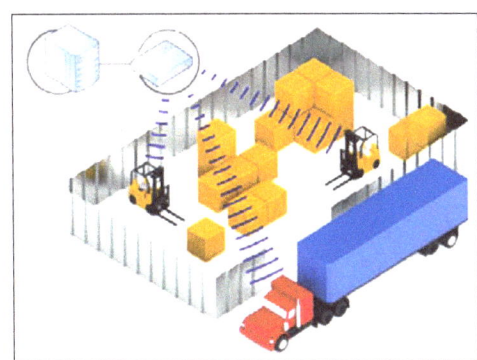
Transport sector M2M apps often involve GPS monitoring (Source: http://aggregate.tibbo.com)

simple fleet management of personal as well as commercial vehicles, maritime cargo shipping management to the more sophisticated surface transport GPS systems that adapt to varying road conditions (such as heavy road congestion ahead in the journey) to suggest substitute routes.

Route optimisation of personal and commercial vehicles results in cutting down unnecessary driving, less wastage of fuel and subsequently lower GHG emissions. BCG & GeSI estimate that M2M apps could save almost 1.9Gt of CO_2 emissions by 2020 in the transportation sector by optimising routes of planes, trains, trucks and ships, and ensuring that people and goods are moved as efficiently as possible.

In case of passenger vehicles, different ride-sharing and personal transport optimisation companies that use M2M are swiftly achieving credibility and market share, thus bringing down the total number of vehicle kilometres travelled.

Better fleet-management systems mean improved vehicle diagnostics, higher vehicle lifespan and lower operating expenditure. M2M can also help in inventory management to bring down losses by eliminating deadheading, for example, return journeys made by empty or underutilised vehicles.

Real estate. By using M2M technologies, it is possible to increase energy efficiency in the built environment without inviting capital expenditure and infrastructure changes.

M2M enables development of extensive and sophisticated building management systems, which monitor various parameters, such as the outside temperature and wind velocity. These systems result in higher security and lower costs, for example, by switching lights and machines on and off automatically. Time is also saved as human intervention is minimised. For example, M2M systems controlling the ventilation and fire systems are set in such a manner that they work optimally at all times.

Architects have started using sensors that capture site-specific data and send it back to the office where the analytic software helps to maximise the potential efficiency of a building in blueprint drafting. M2M can reduce CO_2 emissions by as much as 1.6Gt by 2020 enhancing the energy efficiency of building systems which include heating, cooling, ventilation, lighting, electronics and appliances and security systems.

Agriculture. Lately, we have been seeing deviations in climate and unexpected floods and typhoons. Our agricultural sector is threatened by these and thus demands a change in the manner we grow, harvest, transport and store food.

M2M can help bring down the methane emissions which are produced from cattle digestion (enteric fermentation) by monitoring cattle health and optimising their grazing patterns.

Remote monitoring of soil conditions, smart farming and smart watering are some other ways by which M2M technologies can help reduce GHG emissions. These come under the genre of 'precision agriculture.' For example, tractors that have M2M-enabled auto-steer technologies can be mapped with high precision in terms of their location in the field. They can then automatically reference that data with information on crop aspects, such as soil and water

quality from other sensors and use water and fertilisers very accurately over each square metre of the field with almost zero wastage.

There are also savings in fuel consumption as tractors are used only where and when they are really required. A great example of M2M technology at work in the farms in India is NanoGanesh—a device that enables farmers to switch on irrigation pumps used for watering crops in remote locations using their mobile phones along with a mobile modem that attaches to the starter of the irrigation pump. This is a big boon for the farmers who had to walk several kilometres to their fields to switch on or switch off their water pumps. Nano Ganesh allows farmers to remotely check to see that there is electricity, and to automatically turn the pump on or off.

It is estimated by BCG & GeSI that M2M could help reduce another 1.6Gt of CO_2 emissions by 2020 in the agriculture sector by reducing deforestation, managing livestock and improving the efficiency of planting, seeding, harvesting, fertiliser application and water use. This will enable a larger amount of crops to be grown using fewer resources and in saving money for our farmers.

Endpoint

With time, more and more things in our daily lives will become connected—our home appliances, houses, cars and offices. And the more they will be able to communicate with each other. We can look forward to a whole new and completely connected world.

As per the 2013 report *Machine to Machine Technologies: Unlocking the potential of a \$1 trillion Industry By 2020,* produced by the Carbon War Room Research and Intelligence Group, there will

Some day machines will verbally communicate with humans and other machines (Source: http://get.visagemobile.com)

be 12.5 billion M2M devices globally by 2020. Another transformation that we will see is the shift in the distribution of M2M applications.

In the early days, the M2M apps were mostly centred on the automotive and transport sectors and were commercial in nature. As the technology developed, applications in other sectors, such as health, agriculture and energy, also became prominent. However, their usage was more concentrated in the developed nations.

In time to come, M2M apps will dominate all sectors and include several other sectors, such as banking and real estate. We will have smart homes, smart clothing, smart factories, smart cities and smart anything that you can think of. These apps will be both commercial and consumer in nature and touch several people in emerging and developing countries too.

The popular science-fiction film Star Wars exposed a world where machines verbally communicated with humans and other machines. In the years gone by, it seemed the technological wonders depicted were too far-fetched to ever resemble reality. However, some of these once-futuristic concepts would soon become reality. ●

The author is an associate professor at School of Management Sciences, Apeejay Stya University

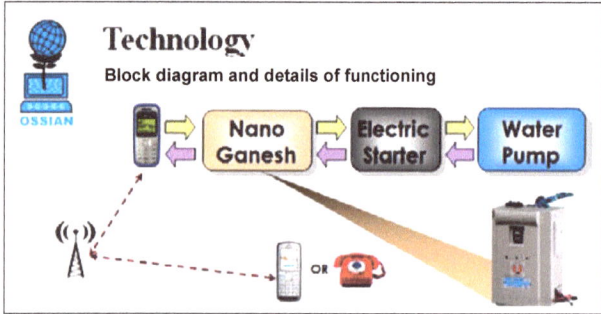

How Nano Ganesh works (Source: http://www.technospot.in)

PIC12F157X 8-bit Microcontrollers

Small Form Factor with High-Resolution 16-bit PWMs

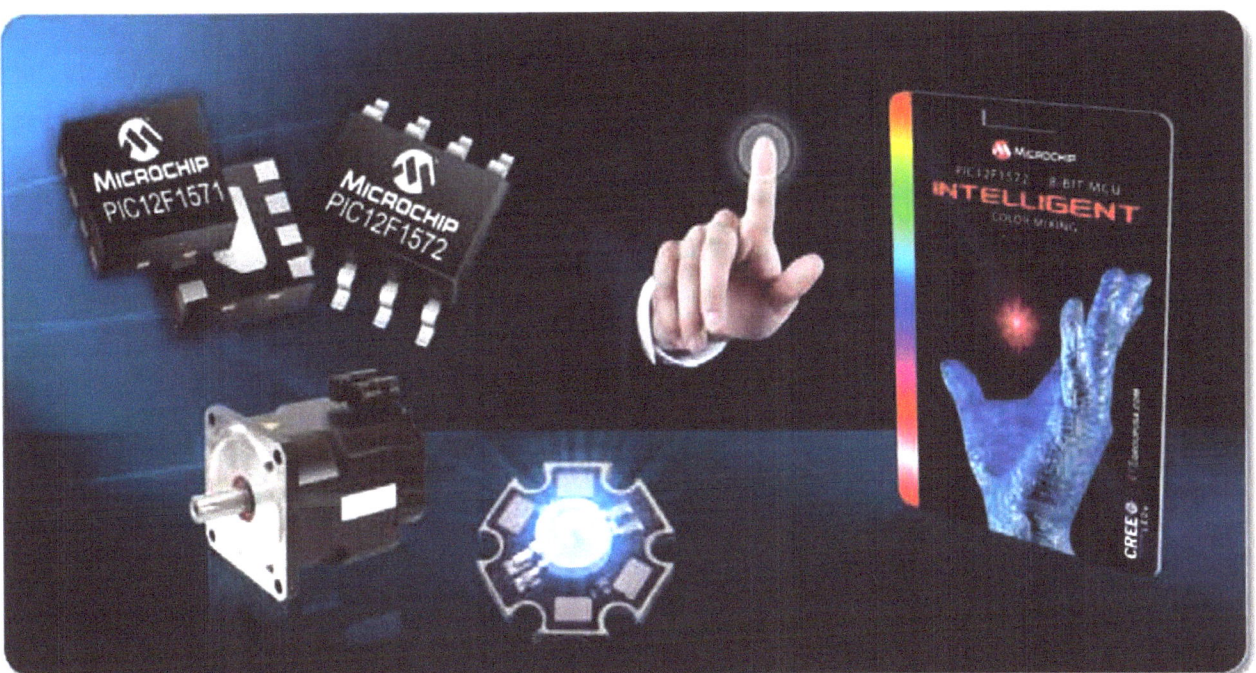

The PIC12F157X family of MCUs combine high-resolution 16-bit PWM drive, closed-loop control and communication capabilities into an 8-pin form factor, enabling increased precision for drive and control in cost-sensitive applications. With operating currents as low as 30 µA/MHz, these products are ideal for lower-power applications. In addition, on-board Intelligent Analog peripherals can be connected internally with the PWM modules to facilitate closed loop feedback without requiring pins or using PCB space, which simplifies your design process.

Microchip's 16-bit PWM Features

- Additional PWM modes
  - Center aligned mode to improve EMI
  - Set/toggle on register match to add flexibility
- Independent timers to drive separate power stages or motors and improve configurability
- Multiple internal compare modes

PIC12F1572 RGB COLOR MIXING DEMONSTRATION

The 8-pin PIC12F1572-based color mixing RGB badge provides a detailed insight into color mixing theory and demonstrates how PIC® microcontrollers can provide intelligent color mixing for LED lighting applications.

www.microchip.com/lighting

India Sales Offices:
Bangalore (080) 3090-4444 • New Delhi (011) 4160-8631 • Pune (020) 3019-1500
Email: asia.inquiry@microchip.com

 MICROCHIP

Microcontrollers • Digital Signal Controllers • Analog • Memory • Wireless

Desktop Manufacturing Equipment

With the market becoming more and more competitive in every segment, time-to-market is emerging as the most important factor defining a product's success. However, designing and prototyping a product takes a long time as each part has to be produced in small quantities at different manufacturing units. With the advent of desktop manufacturing, most of these parts can now be produced in-house, saving a lot of time

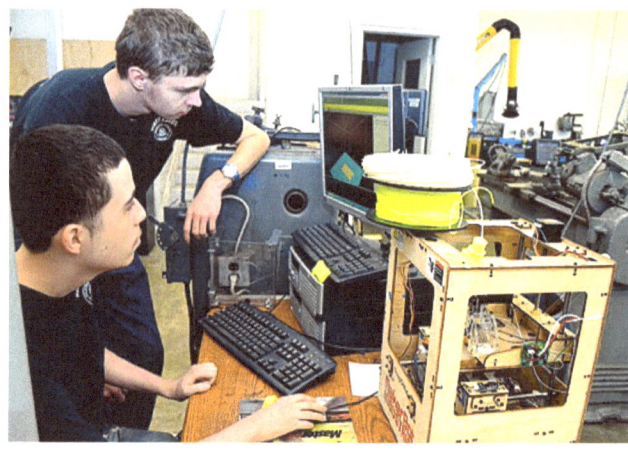

ANKIT GUPTA

Designing and prototyping a product takes a long time, not because of the complexity involved, but because of the lead times associated with the outsourced production of each part. Most of the outsourced processes are sequential in nature, so the overall lead time for prototyping becomes the sum of individual lead times for each process.

For a new product, the first thing done is mechanical design freeze. All other components, such as assembled PCB boards, are designed to fit into the decided mechanical design. So everything has to wait until the mechanical design and prototyping is complete. Mechanical designs also get revised later, if other parts have some constraints.

Once the mechanics is done, PCB designs are finalized and their gerber files are sent to the PCB manufacturer. These files have to be revised sometimes to suit capabilities of the man-ufacturer. For more advance PCBs, with track width below 4mil, you may have to get the work done abroad. As the manufactured PCBs are sent to you via courier, you should include this trans-portation time also in your time schedule.

The PCBs and the components to be mounted on them are then sent to an EMS (electron-ics manufacturing service). Fortunate-ly, components procurement doesn't have to be sequential; you can do it in parallel with production of the PCBs or earlier. Some EMS can even procure the components themselves, but they are reluctant for low quantities. So you may have to procure them yourself.

If a PCB board has SMD compo-nents, the EMS will have to make the stencils for automatic placement of the components through chip shooters. But you can ask them to do it manu-ally, if the quantities are small. Again, the work will be done depending on the availability of resources and the process can take weeks.

We have not included the negotia-tion time taken with all these vendors after the receipt of their quotes.

When the boards are ready, fitment in the mechanics is checked and then the boards are tested for functionality and compliance. If there happens to be some flaw in design or non-compliance with the standards and specs, the whole process will have to be repeated.

Some manufacturers provide turn-key solutions, where they make the PCBs, source the components, make stencils and assemble PCBs, all at one place. Though this saves some time, the processes usually need to be much faster to meet the pace of the market. Some of the key issues involved in producing the prototypes are:

Low quantities, low priority. For prototyping you would only be producing a small batch of, say, ten pieces. If you see the business angle, it will not interest most vendors, or the work will be done at a much higher cost, which can hit your pro-ject's budget. The components you buy from online stores will come at very high prices, and your work at EMS will be done on low priority due to their small quantity.

Tooling cost for all parts. For pro-ducing the PCBs and assembling them at EMS, you need to pay an additional one-time cost, which is called tooling or setup cost, for every project. And if the design gets revised after first pro-totyping, which is normally the case, you will have to pay the tooling cost again. Also, if there is a change in PCB design, new stencils will have to be made and the money spent on previ-ous stencils will go waste.

Time delay. As already mentioned, a lot of time is wasted usually in cost negotiations, transportation, sequential nature of processes and delays caused at EMS due to lesser quantities, other than the processes themselves.

With the recent availability of desktop manufacturing equipments, such as SMD pick-and-place machines,

TABLE I
DIY-Type 3D Printers

Vendor	Product
3D Printronics	Felix 2.0
Lulzbot	TAZ 4
RS Components	RepRap Pro omerod
Chipmax	MBot Cube
Amptronics	Prusa Mendel LM8UU

TABLE II
Assembled Desktop
3D Printers

Vendor	Product
3D Systems (formerly ZCorp)	Zprinter 850
Afinia	H479
3D Protomaker	Sprint
Makerbot	Replicator 2
Airwolf 3D	AW3D HD
Altem Technologies	uPrint SE
Amptronics Systems	Dual Extruder 3D Printer
Brahma 3	Brahma3 Anvil
CAD Centre	ProJet 7000 HD
Cubify	CubeX
Formlabs	Form 1
J Group Robotics	Dimension LE
KCBots	KUBE
Stratasys	Objet 1000

Fig. 1: A 3D printer

ovens, PCB-prototyping machines, stencil printers and 3D printers, all the above-mentioned prototyping work can now be done in-house and a lot of time and money can be saved. These equipments are expensive but, in the long run, can save you money and, most importantly, the time to market. Let us have a look at these equipments one by one.

3D printers

Using 3D printers, which got introduced commercially only recently, you can produce almost any part. These printers can be used for checking the mechanical design and fitment. In regular process, you get the designs made and then send them to the manufacturer. Then there is a long wait be-

fore you receive the prototype to check everything and finalise the mechanics. With a 3D printer you can immediately fabricate your design and check it. It accepts .stl files and converts them to gcode files to print all layers one by one. Fig. 1 shows a 3D printer.

The material of the printed prototype will off course be different, but you only want to check the dimensions, fitment and integration of different parts. Sometimes your prototype could have dimensions larger than what your 3D printer can handle. In that case you can scale down the dimensions of each part and check the fitment and integration in this scaled down dummy.

3D printers also come in the form of DIY kits, such as Prusa Mandel i2. You get all the parts of the printer but you need to follow its manual to assemble it and install everything before you can produce anything. These printers are normally less precise and you have to do a lot of calibration before starting work on them. But such printers are usually cheaper and can be good for basic mechanical models' dummy checking. Table I lists some manufacturers of the DIY-type 3D printers.

But if you are looking for more precise and reliable prototyping, you should go for ready-to-use 3D printers. These printers come assembled with reliable mechanical structure and give precise results as per specs mentioned by their manufacturers. Most of all, nil or very little calibration is required in such printers. Table II shows some assembled desktop 3D printers from different manufacturers.

Here are some more pointers to

help you select the most suitable 3D printer for you:

Print platform. This is the maximum dimension that a printer can print. Select appropriate dimensions for your requirement. If the prototype is larger than that, either print a scaled down version or break it into parts that can later be combined.

Resolution. Check for horizontal and vertical resolutions in the specs sheet. The vertical resolution will define the thickness of each layer. Better resolution prints smoother surfaces. Similarly, horizontal resolution defines how fine the extruder can move in XY plane. The printer with finer XY resolution can print finer features in the model.

Print speed. This will give you an idea about how fast your designs can be printed. It is normally defined as the time it takes to print a specific distance in the Z-axis.

Number of extruders. Extruders are the print heads. The number of extruders is generally associated with the number colours that the printer can handle at a time.

SD card and display support. Some printers have an interactive display to help select all the functions, but this comes at an additional cost. SD card support is also an important feature; the design can be copied in the SD card and the printer keeps printing without the need of an attached computer.

Desktop PCB prototyping

With the mechanical design verified, the next step is to make your PCB design such that it fits into the mechanics. Once the design is done using PCB design software, within the dimensional constraints set by the mechanics, you will have to get some of these PCBs made so that you can mount all the components and test the boards for functionality.

Since most manufacturers are not interested in low quantities, it is best to make these PCBs in-house. Though some people use photo-resist and UV systems for making a few pieces, these techniques are inaccurate and time

TABLE III
PCB Prototyping Machines from Different Manufacturers

Manufacturer	Product	System
MitsPCB	FP21T	Mechanical
AccurateCNC	Accurate 426	Mechanical
Colinbus	CPR series , PCB box	Mechanical
T-Tech	QC series	Mechanical
LPKF Laser and Electronics	ProtoMat S103	Laser
Everprecision Tech Co	EP-2006	Laser

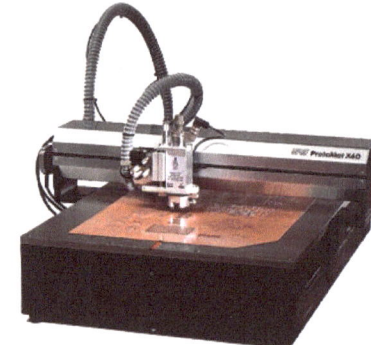

Fig. 2: A PCB-prototyping machine

TABLE IV
SMT Pick-and-Place Machines

Manufacturer	Products
LPKF Laser and Electronics	Protoplace S
Mechatronics Systems	M1, P1
SMTnet	Manncorp 7700-FV Auto Placer, Manncorp SMT 2000 (manual)
Essemtec	Paraquda, Pantera-X, Cobra

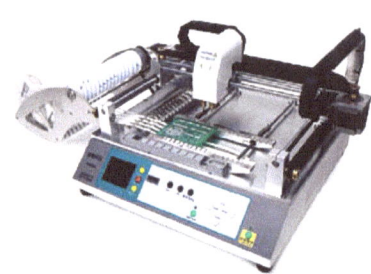

Fig. 3: A desktop SMT pick-and-place machine

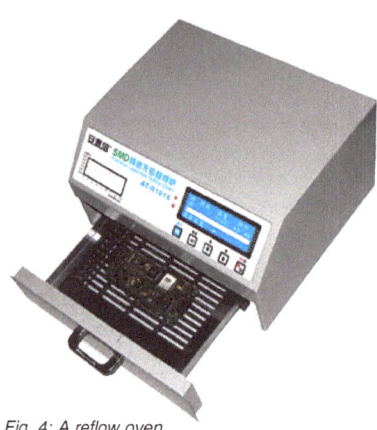

Fig. 4: A reflow oven

consuming. Besides, making double-layer PCBs with such techniques is very cumbersome.

Various desktop PCB-prototyping systems available these days help in making the PCB prototypes very quickly (refer Fig. 2). These systems are majorly of two types: mechanical and laser milling systems. Both types of systems create conductive paths and pads by milling insulating paths. The insulating paths separate the electro-conductive copper surfaces that form the network of conductive paths.

A mechanical milling system contains a rotating drill head that works on the copper-clad sheet, milling all insulating paths. It is comparatively slower than a laser system which uses an energy-emission device to focus a highly concentrated stream of photons onto a small area of a work-piece for milling the insulating paths. Table III shows some mechanical and laser-type PCB-prototyping systems from different manufacturers.

Next, all the required holes need to be drilled. Both these systems can drill holes for mounting through-hole components. The specifications of these systems are simple and easy to understand. You can look at specs, such as working area, resolution, speed, drill sizes and power consumption, to decide a suitable system for you.

SMT pick and place

Once the PCBs are ready, you need to mount all the components and test the boards. Mounting of through-hole components is easy but SMD components require some skill and patience. So most people avoid doing it themselves and outsource the job, which costs money for making stencils and tooling, and time delay is always there.

With desktop SMT machines the SMD boards can be assembled in-house. Fig. 3 shows a desktop SMT pick-and-place machine. Such machines come in completely automatic and semi-automatic forms.

In an automatic system, no operator intervention is required. The system automatically applies solder paste on the pad and then places the components on corresponding solder beds. The placement is not very accurate in most cases, but during reflow oven stage the parts tend to organise nicely.

In a manual system, the operator controls the application of solder paste and placement of parts. The placement can be much more accurate in this case. But you may not want an operator, who could make errors. Besides, the process will be much slower than with an automatic machine.

Some manufacturers have also come up with solutions that include PCB milling, solder-paste application and pick and place, all in one system.

TABLE V
Reflow Ovens from Different Manufacturers

Manufacturer	Product
LPKF Laser and Electronics	ProtoFlow S
T- tech	SMT reflow oven –A
Mechatronics Systems	RK320 , RK360
SMTnet	Manncorp 850 Benchtop, Manncorp BT300CP Benchtop
Essemtec	RO-06-PLUS

Do check the specs of all the three systems separately to see if they suit your requirements. Table IV shows some desktop pick-and-place machines from different manufacturers.

Reflow ovens and wave-soldering machines

The solder paste is applied and components are placed on the solder beds by the pick-and-place machines. But these components are not soldered to the

Fig. 5: A wave-soldering machine

pads as yet. For that you need to put these boards in an oven that heats up the boards and melts the solder paste. The melted solder paste sticks the pins of the components to the pads when it cools down.

Ovens also come in automatic and manual forms. In an automatic system, you decide the environment setting and put the board in the oven. The board ejects after the required time automatically. In a manual oven, you set the temperature and put the board in and take it out yourself after an estimated time period. Fig. 4 shows

a reflow oven and Table V lists some ovens from different manufacturers.

Though most people prefer to mount through-hole components manually but, for faster turnaround, a desktop wave-soldering machine can be used. These machines have melted solder in a flat bed, and the boards are moved over it slowly so the through-hole solder points catch some solder. The solder sticks the pins with the pads, making a permanent joint on cooling. Fig. 5 shows a desktop wave-soldering machine.

With all these equipment available, you can reduce the prototype-development time drastically. This, in turn, can reduce the time to market, increasing your product's chances of being a success. If you do not want to spend money on all these equipment in one go, select those that you really need to increase the efficiency of your product development cycle. ●

The author is a technical editor at EFY

Part 3 of 4

Defence Lasers and Optronic Systems:
Semiconductor Diode Laser Electronics

Semiconductor diode lasers are close second to solid-state lasers in usage as far as tactical military applications of lasers are concerned. Laser aiming modules fitted on small arms and used to enhance night-fighting capability of infantry soldiers, short-to-medium range eye-safe laser rangefinders used for observation and surveillance, laser proximity sensors and short-range laser dazzlers are some prominent examples. Focus in this part of the article is on semiconductor diode laser electronics

DR ANIL K. MAINI AND NAKUL MAINI

While semiconductor diode lasers, or simply laser diodes, are extensively used in a range of laser devices intended for tactical military applications, such as short and medium-range laser rangefinders, proximity sensors, short-range laser dazzlers and laser aiming devices, it needs to be emphasised here that they are invariably used as the optical pumping source for all Nd:YAG laser-based military systems.

Laser-diode electronics primarily comprises a constant-current source that can provide to the forward-biased laser diode the desired magnitude of current and also has in-built features to provide protection to the device against all those parameters it is adversely sensitive to. The other important circuit block is the temperature controller that can maintain the laser diode junction at the desired value of constant temperature irrespective of ambient temperature. The need for a precise constant-current source and constant-temperature operation arises from the dependence of laser-diode wavelength on drive current and operating temperature. Not all laser-diode applications, including military applications, have stringent requirements of ultra-stable wavelength necessitating a high degree of current and temperature stabilisation, though for efficient and reliable operation, stability of

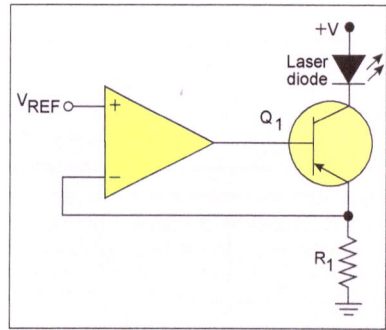

Fig. 1: Constant-current mode

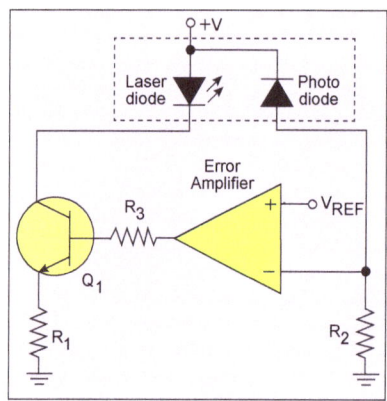

Fig. 2: Constant-output-power mode

drive current and operating temperature are always desirable.

Although laser diodes exhibit excellent reliability under ideal operating conditions, these are highly susceptible to damage due to excessive drive current, electrostatic discharge (ESD) and transients. Laser diode damage manifests itself in the form of one or more of the following prob-

lems, namely, reduced output power, shift in threshold current and increase in beam divergence and failure to laser action, thus producing an LED-like output only.

Relevant damage mechanisms are overcurrent, overheating, current spikes and power-line surges. Improper grounding and shielding lead to radiated electrical transients. Improper power-line conditioning also leads to severe fast transients. ESD caused by improper handling during storage, transport and mounting is considered as the single leading cause of premature laser diode failure.

Constant-current and constant-power operational modes

There can be two possible circuit topologies the laser diodes can be operated in, namely, constant-current and constant-output power modes. Operating temperature is of course an important factor in both the modes. In view of dependence of laser diodes' performance on operating temperature and drive current, the preferred mode of their operation is the constant drive current with precise control of operating temperature. Constant-current operation without temperature control is generally not desirable. Even at constant drive current, output power would increase with decrease in operating temperature. With significant decrease in temperature, output power could easily go past the absolute maximum value.

The circuit topology in the case of constant-current mode of operation is usually configured around a current-sensing element that continuously senses the drive current and produces a proportional voltage. This is then compared with a reference voltage representing the desired value of drive current to generate an error signal. The error signal, after suitable conditioning, is fed back to restore the drive current to the desired value. Fig. 1 shows the basic schematic arrangement of constant-current drive circuit.

Practical circuits with in-built protection features are far more complex. Typical constant-current drive circuits and related design issues are discussed in the following paragraphs.

There are applications where constant output power is more relevant. In such a situation, direct dependence of output power on drive current can be exploited to maintain a constant output power. Even when a laser diode is driven by a constant current, the heat dissipated at the laser-diode junction leads to rise in temperature and hence fall in output power. Increase in temperature could be because of absence of any active temperature control mechanism or even inadequacy of heat sink in the presence of active temperature control. This reduction in output power can be compensated by increasing the drive current. The current source can be designed in such a way that it adjusts the drive current in a feedback mode to maintain a constant output power instead of maintaining a constant drive current. Fig. 2 shows the basic circuit topology. The basic circuit topology of constant-output-power drive circuit is similar to that of constant-current drive circuit, except for the nature of sense signal, which in the present case is a photo current proportional to the output power.

Constant-output-power drive circuit needs to be operated with an absolute current limit to prevent any

Fig. 3: Laser diode with integral photodiode

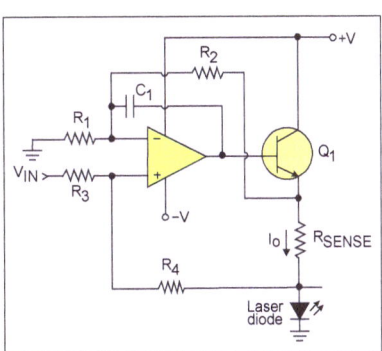

Fig. 4: Constant-current laser diode drive circuit for grounded load

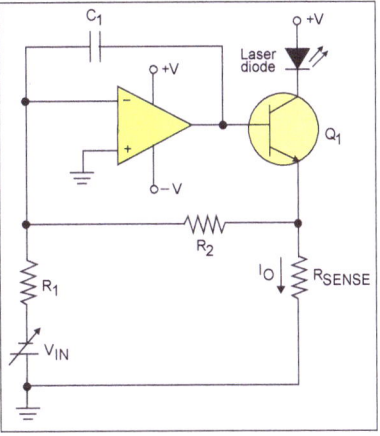

Fig. 5: Constant-current laser diode drive circuit for floating load

thermal runaway problem caused by uncontrolled increase in drive current. Laser diode modules with an integral photodiode (Fig. 3) facilitate constant-output-power operation, though noise intrinsic to the integral photodiode manifests itself in the form of noisy and unstable output in the case of constant-output-power mode of operation. The device usu-

ally has three terminals, including either anode or cathode of laser diode, either anode or cathode of photodiode and a common terminal obtained by connecting the remaining terminals of laser diode and photodiode. In some cases, all four terminals, two of laser diode and two of photodiode, are brought out.

Laser-diode drive circuit: constant-current mode

Fig. 4 shows the basic constant-current source circuit for grounded-laser-diode configuration. Laser-diode current is sensed differentially by measuring voltage across R_{SENSE} wired in series with laser diode. Laser diode current in this case can be computed from $I_O = V_{IN} \times R_4 / (R_3 \times R_{SENSE})$, where $R_1 = R_3$ and $R_2 = R_4$. C_1 is the compensation capacitor connected for stable operation. V_{IN} may be derived from a band-gap reference. In case the control voltage V_{IN} is of negative polarity, it is connected to R_1 and R_3 and is grounded instead. A digitally-controlled current source may use a voltage output digital-to-analogue converter to generate V_{IN}.

Similar circuit for a floating load is shown in Fig. 5. Laser-diode current in this case can be computed from $I_O = V_{IN} \times R_2 / (R_1 \times R_{SENSE})$.

The voltage-controlled constant-current sources discussed so far provide constant-current drive to the laser diode, provided the DC supply voltage, the voltage at the emitter terminal and the resistance in the emitter lead were all constant. It would be a reasonably good assumption if the source voltage were derived from a precision band-gap reference and the resistors used had stability specifications equal to or better than the desired level of current stability. But when it comes to achieving a higher level of current stability, say ±10ppm or better, a feedback loop that in situ samples the diode current and applies a correction in the case of drift in drive current becomes essential. A negative feedback loop of this kind would reduce the drift or error in the drive current by a factor that equals the loop gain.

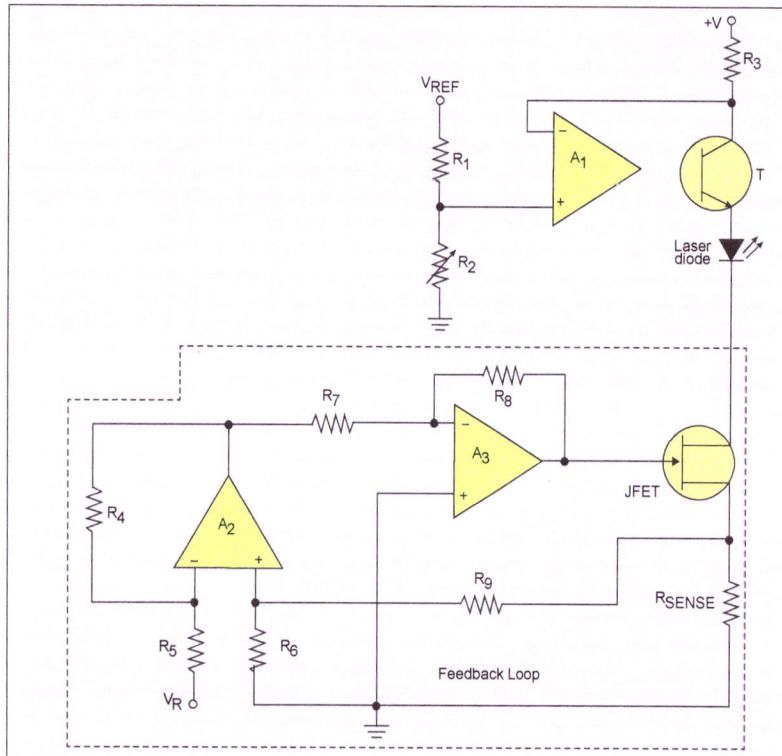

Fig. 6: Laser-diode precise current control

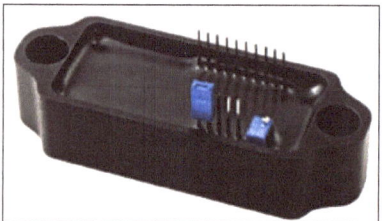

Fig. 7: Precision constant-current laser diode driver for OEM applications

One such circuit is shown in Fig. 6. The basic circuit topology is similar to the one used in driver circuits described in the previous pages. The only change is inclusion of a junction FET (JFET) connected in series with the sense resistor and wired as a voltage variable resistor. A small variation in drive current is compensated by an appropriate variation in the drain-source resistance of the JFET.

Initially, at the nominal value of drive current, circuit parameters are so adjusted as to ensure that JFET with the feedback loop closed gets a negative gate voltage to keep it nearly in the middle of its VVR characteristics. This is done to fully exploit the voltage-

dependent resistance range. Precision constant-current laser diode drivers

are commercially available today from a host of manufacturers for original equipment manufacturer (OEM) applications. A representative photograph is shown in Fig. 7.

Protection features, such as slow start, immunity to fast transients and overcurrent limit, as outlined earlier, are essential features for every laser diode driver circuit to have. Fig. 8 shows a constant-current laser diode drive circuit with modulation capability and in-built above-mentioned protection features.

Let us first talk about protection features provided by the drive circuit. The Schmitt comparator at the input provides a delay of the order of a few tens of milliseconds after the switch-on to offer protection against switch-on transients. The time delay is decided by the R_1C_1 time constant. R_2 and C_2 provide the desired slow or soft start and decay during switch on and switch off. R_2C_2 time constant is of the order of one to two seconds. R_{11} and C_3 provide additional protection against transients; D_1 protects the laser diode against reverse voltages.

The constant value of drive current is decided by the voltage present at the noninverting input of op-amp A_4, which equals the voltage present at the output of op-amp A_2, provided R_8 = R_9. This is further equal to sum of voltage levels due to modulation input and reference voltage V_R after potential divider

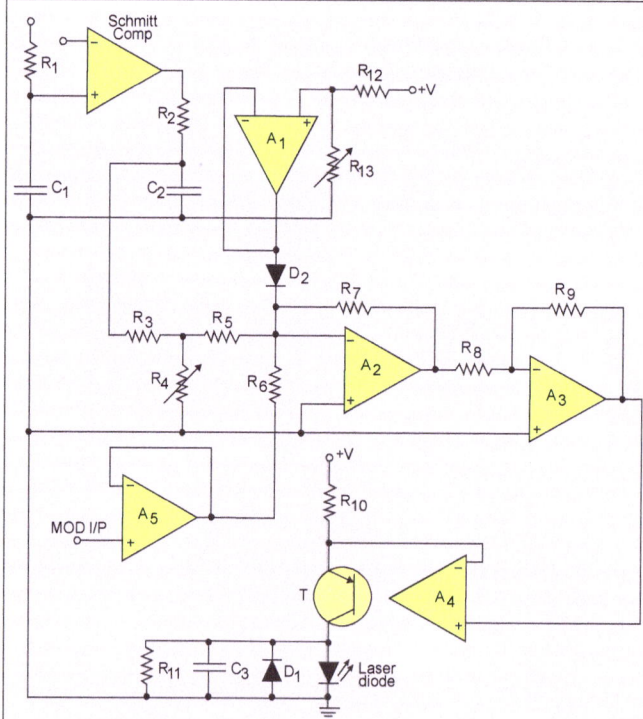

Fig. 8: Constant-current drive circuit with protection features

INNOVATOR IN ELECTRONICS

**Discovering the Best Solutions
For Every Electronics Challenge**

One-Stop Solution

Murata has helped enable numerous innovations
in countless applications throughout the communications, automotive,
healthcare, environmental and energy areas.
Apart from being recognized for our product reliability and
our total solution capability, Murata can provide you with the advantages
of quick response times supported by our strong global network.
Murata Electronics - the one-stop that fulfills
your electronics needs across the globe.

Superior Quality Products

From standard electronic passive
components to communication modules
to power, Murata provides a comprehensive
product offering with reliable quality.

Technical Support

Murata's expertise and technical support
resources provides you with total support,
all the way from the design stage, to trial
runs and to mass production.

Global Network

Murata has teams of sales personnel and
engineers located in every part of the
world. Our strong network can support all
your activities - design, procurement, mass
production, on a local and global level.

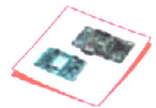

Ceramic Capacitors

Inductors /
EMI Suppression Filters

Thermistors /
Variable Resistors

Sensors

High Frequency Devices /
Communication Modules

Power Modules

Murata Manufacturing Co., Ltd.   http://www.murata.com

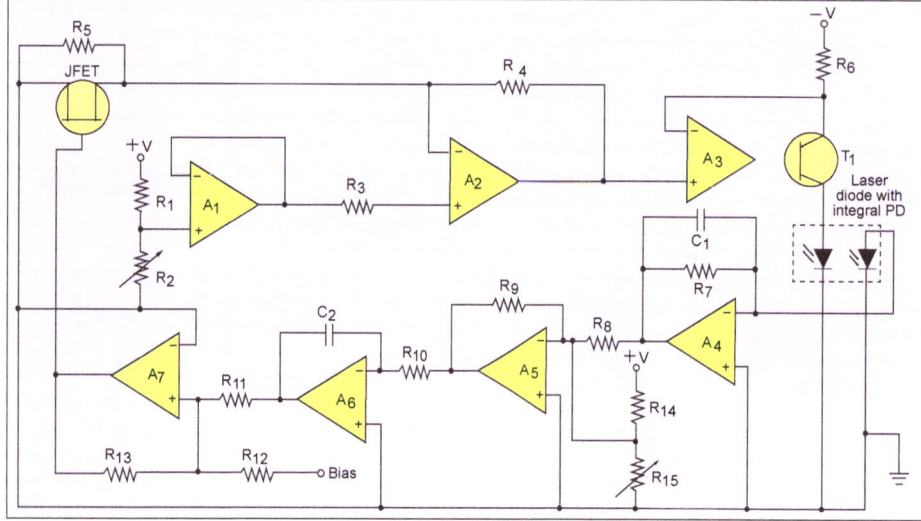

Fig. 9: Laser-diode drive circuit for constant-output power

constituted by R_3 and R_4 and also assuming that $R_5 = R_6 = R_7$. This is true only if voltage V after potential divider arrangement R_{13} and R_{14} and appearing at output of op-amp A_1 did not interfere. This is true as long as diode D_2 is reverse biased, which it is during normal operation.

If the drive current exceeds a certain preset limit governed by voltage at the output of op-amp A_1, diode D_2 gets forward biased and clamps the voltage at the noninverting input of op-amp A_4 to the voltage at the output of op-amp of A_1, thereby providing overcurrent limit. The circuit may be modified to include a JFET and associated feedback loop components as shown earlier in Fig. 6 to improve current stability.

Laser-diode drive circuit: constant-output-power mode

Laser diode can be driven to maintain a constant-output power by varying the drive current in accordance with the output power. The feedback loop circuit is such that it increases or decreases the drive current in response to decrease or increase in output power. Fig. 9 shows the constant-output-power drive circuit. The integral photodiode produces current proportional to optical output power from the laser diode. This photocurrent is converted

into a proportional voltage using a transimpedance amplifier configured around op-amp A_4.

The voltage representative of laser power is then summed up with a reference voltage representing the desired power level in an inverting summer configuration A_5. The output of the summer, which is null when the actual power level equals the desired power level, feeds integrator A_6 that provides the correction signal even for infinitesimally small deviations from the desired power level. The integrator output summed up with a bias voltage feeds control element, which is a junction FET in this case.

The drive current, and hence the output power, is governed by voltage present at noninverting input of op-amp A_3. Evidently, reduction in laser-diode-output power causes reduction in photodiode current, which finally leads to gate terminal of JFET becoming more negative. This further causes a reduction in voltage at A_3 noninverting input, increasing the drive current to restore the output power. Similarly, increase in laser-diode power reduces drive current to restore the power at nominal value.

Driving laser diode in pulsed mode

There are two possible modes of operation of laser diodes to produce a

pulsed output. In one of the operational modes of relevance to their use in military devices, such as laser rangefinders employing time-of-flight principle and laser proximity sensors, the laser diode is driven by current pulses that are a few tens to a few hundreds of nano-seconds wide. In the other operational mode, called quasi-CW mode, laser diode is driven by current pulses that are typically hundreds of microseconds to a few milliseconds wide. This operational mode is invariably used in the case of laser-diode arrays pumping solid-state lasers, including those for military rangefinders, target designators, electro-optic countermeasures and so on.

Quasi-CW operation of laser diodes for optical pumping of solid-state lasers is at relatively low repetition rates of typically a few hertz to a few tens of hertz. This allows them to be operated at relatively high peak powers, which is made possible due to low duty cycle of operation of quasi-CW devices, which keeps the average power low.

An important consideration while designing laser-diode drive circuits for pulsed operation, conventional or quasi-CW operation, the laser diode(s) must not be switched between cut-off and the nominal maximum value. This is highly detrimental to the life of laser diodes. They must always be operated between lower values slightly greater than the lasing threshold and the nominal value. Laser diode driver circuits for conventional pulsed and quasi-CW operation are configured around the same basic building blocks as those described earlier in the case of CW operation. The laser diode drive circuit of Fig. 8 can be used for operation in conventional pulsed mode by applying the desired pulsed waveform at the modulation input. The lower value of drive current is governed by

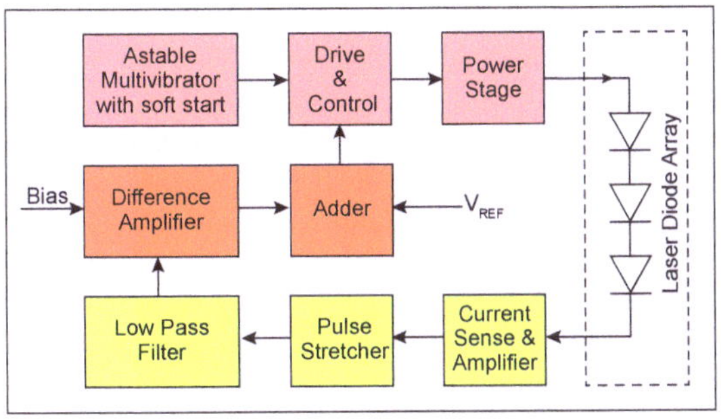

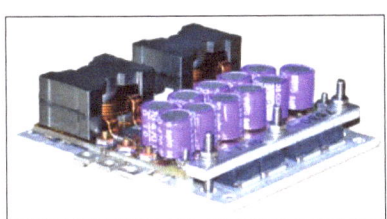

Fig. 10: Laser diode driver for quasi-CW operation

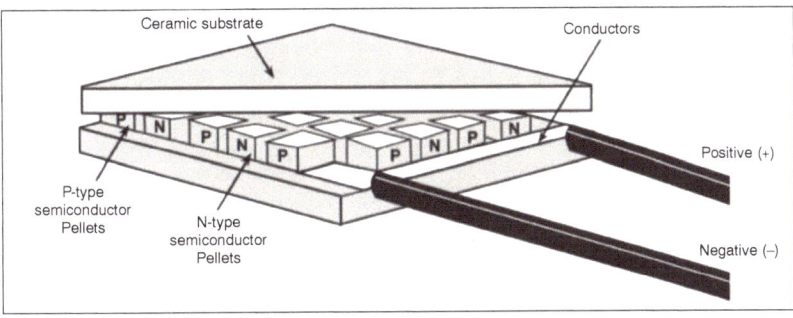

Fig. 11: Quasi-CW laser diode driver

Fig. 12: TE module construction

R_3-R_4 potential divider.

In the case of quasi-CW operation of laser diode arrays used for optical pumping of solid-state lasers, peak-current-pulse amplitude typically varies from a few tens to a few hundreds of amperes. The current drive stage in this case is usually a cascade arrangement of a drive and control stage and a power stage. Drive and control stage is a preamplifier stage with some kind of control feature that allows to vary the amplifier gain and hence the drive current to the power stage. Fig. 10 shows the block schematic arrangement of a typical quasi-CW laser-diode drive circuit.

The circuit operates as follows. The astable multivibrator produces a train of pulses having pulse width and repetition rate at which the laser diode or laser-diode array

needs to be driven. Soft start in-built into the circuit provides protection against damage due to instant power on. Power stage contains the active device(s) and associated circuitry capable of delivering required current pulse amplitude.

The drive and control stage could be configured either around discrete bipolar transistors/metal-oxide-semiconductor field-effect transistors (MOSFETs) or an op-amp with JFET as one of the gain determining elements. The power stage could be configured around one or parallel connection of more than one bipolar transistors

or MOSFETs. The sensing element is usually a resistor connected in series with the diode or diode array. The peak pulse voltage across the sense resistor representing the peak amplitude of current pulse is amplified and then subsequently stretched. The stretched pulse train, after filtering, is fed to a differential amplifier stage where it is compared with a standard reference to produce an error voltage. The error voltage here represents deviation of peak amplitude of current pulse from the desired value. The error voltage is added to a fixed bias and fed to the control element in the drive stage to control the drive current to the power stage.

Again, a large number of manufacturers internationally offer compact laser-diode drivers for OEM market catering to the requirements of CW,

conventional pulsed and quasi-CW operational modes. Fig. 11 shows photograph of one such laser-diode driver for quasi-CW operation. One cannot miss the bank of capacitors connected across the DC supply input to provide high peak current drive capability required in such an operation.

Laser diode temperature control circuits

Laser diode junction gets heated due to power dissipated at the junction, and one of the methods to remove heat away from the junction is by using a heat sink. This passive technique of heat removal using heat sink becomes impractical for moderate-to-large power-diode lasers. Also, it cannot be used to operate the laser diode at a temperature lower than the ambient temperature, nor can it be used to stabilise the temperature. Use of an active temperature stabilisation mechanism therefore becomes essential while working with moderate-to-high-power-laser diodes and also in low-power laser diodes where application demands precise temperature control. Thermoelectric (TE) cooling device based on Peltier effect is the heart of such a system. A practical TE module is usually a two-dimensional array of P-N couples connected electrically in series and sandwiched between two thermally conducting and electrically insulating faces (Fig. 12). Both single-stage and multiple-stage TE cooler modules (Fig. 13) are commercially available. Multiple-stage modules offer higher cooling/heating capacity and maximum differential temperature specifications.

TE-cooler-based active temperature stabilisation circuit uses TE module as the control element, a temperature sensor and a properly designed feedback loop. The TE cooler is a reversible solid-state heat pump whose operation is based on the Peltier effect. According to the Peltier effect, when electric current is passed through a junction of dissimilar metals, heat is created or absorbed at the junction, depending upon the direction of flow of current.

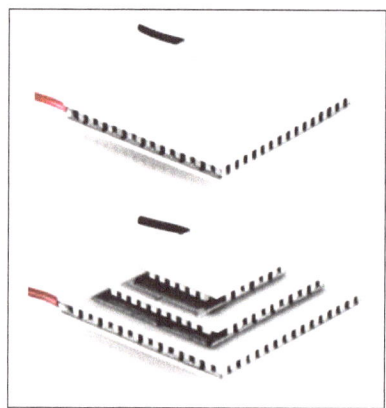

Fig. 13: Single-stage and multi-stage TE modules

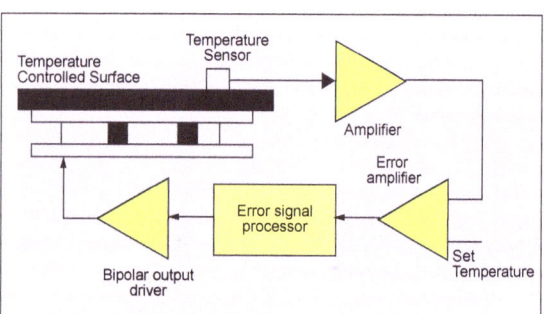

Fig. 14: Schematic arrangement of TE temperature-control circuit

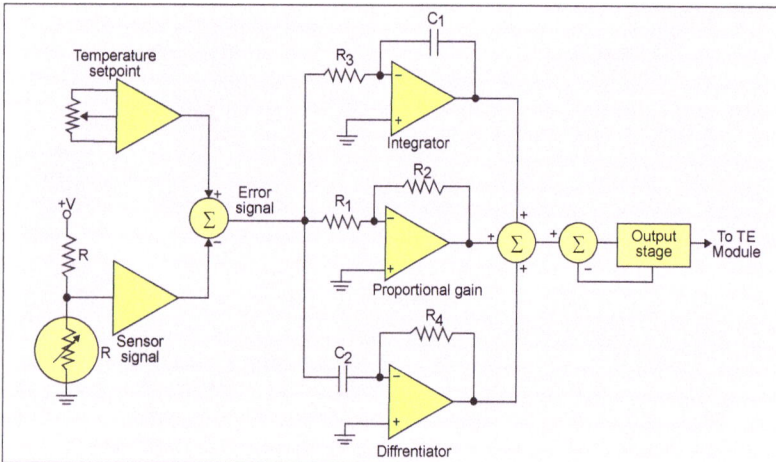

Fig. 15: PID controller and other building blocks

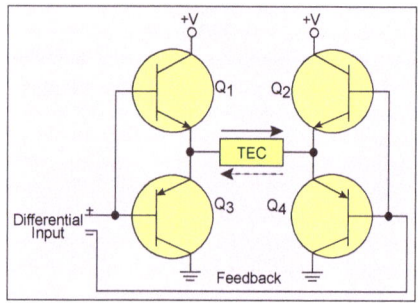

Fig. 16: Output stage configured around half-bridge circuit

The heat transfer according to the Peltier effect takes place in the direction of flow of charge carriers.

A TE module literally pumps heat from one place to another, heating one face and cooling the other in the process. Whether a particular face becomes hot or cold depends upon the direction of current flow through the TE module. Thus, by controlling the magnitude and direction of the current drive to the TE module, it can be used as a control element in a temperature stabilisation circuit. The TE module pumps the heat dissipated in the laser diode on to a heat sink mounted on the supposedly hot face of the TE module. The heat is transferred from heat sink to the ambient.

TE modules are characterised by maximal performance parameters, namely, maximum temperature difference ΔT_{MAX} at zero heat load, maximum heating capacity Q_{MAX} for zero temperature difference, device current I_{MAX} measured at ΔT_{MAX}, terminal voltage V_{MAX} corresponding to I_{MAX} with no heat load and coefficient of performance (COP), which is the ratio of pumped heat load to electrical power supplied to the device.

Fig. 14 shows the basic block

schematic of TE-module-based temperature control circuit. Key building blocks of the drive and control circuit include temperature sensor, error amplifier, error signal processor and a bipolar output drive circuit. Bipolar output driver provides the required power to drive the TE module. The output stage in most cases is designed to allow flow of drive current in either direction through the TE module. This enables both cooling and heating of the device to maintain its temperature at the specified value, regardless of ambient temperature being higher or lower.

Negative temperature coefficient (NTC) thermistor is the most commonly used temperature sensor due to its high sensitivity. RTD and semiconductor sensors have better linearity but suffer from poor sensitivity. Semiconductor sensors are also available in integrated circuit form producing a current linearly related to absolute temperature. AD 590 is one such sensor. Balanced bridge circuit configuration that generates a differential output is generally used. Error amplifier is configured around an op-amp. Error-signal processor could be anything from simple on-off controller to a proportional controller or proportional-integral (PI) or even fully digital PI-differential (PID) controller.

On-off controller is never used in practice. In proportional control, the drive signal to the TE module is proportional to the difference between the actual and desired temperatures. In a proportional controller, there is always a residual error even after the controller has settled to the final state. This error is proportional to the difference between the desired and the actual temperature and is inversely proportional to the gain of the control loop.

The problem of residual error of a proportional controller can be overcome by the addition of an integrator in the control loop. The result is a PI controller. One disadvantage of PI control is that it would be slow to respond to large residual errors. PID overcomes this problem encountered in PI controllers. Addition of derivative term improves the loop's transient response. This type of controller is mainly used in applications where large thermal loads must be controlled rapidly and accurately. Fig. 15 shows a typical PID controller interfaced with other building blocks, including temperature sensor, reference generating circuit and error amplifier.

The output stage provides necessary drive power to the TE module. In most cases, electronic systems are designed to operate from a single positive DC voltage supply. Also, in most applications, TE modules need to be driven in bipolar mode to cater for both heating and cooling operations. A commonly used circuit topology to provide bipolar drive to TE modules while operating from a single DC voltage is the half-bridge circuit topology (Fig. 16). The circuit is usually driven at the input by a

Fig. 17: Integrated laser-diode drive and temperature controller

driver amplifier stage with a differential output. Transistors Q1-Q4 and Q2-Q3 conduct alternately to provide bidirectional operation of TE module. A cascade connection of two half-bridge circuits may be used to enhance voltage and current drive capability of the output stage.

TE drive and control modules for laser diode temperature control and stabilisation are commercially available, both as general-purpose benchtop equipment as well as for OEM applications. Even integrated-laser-diode drive and temperature-control modules are available for OEM applica-

tions. Fig. 17 shows a representative photograph.

Present-day military laser rangefinders and target designators are largely configured around diodepumped pulsed solid-state lasers. We have discussed in the preceding paragraphs the building blocks of laser-diode drive and temperature-control electronics. Circuit topologies of laser-diode drivers for different modes of operation of laser diodes, including CW, quasi-CW and pulsed modes and also laser-diode temperature drive and stabilisation, have been discussed. Gas laser electronics is in focus in the concluding part of the article.

To be continued

Dr Anil Kumar Maini is a senior scientist, currently the director of Laser Science and Technology Centre, a premier laser and optoelectronics research and development laboratory of Defence Research and Development Organisation of Ministry of Defence. Nakul Maini is a technical editor with Wiley India Pvt Ltd

What is Inside This Month's DVD

This month's DVD includes an EDA package with a professional standard printed circuit board design and layout tool. If you are a programmer, we have also included handy HDL simulators and an Arduino IDE

PANKAJ V.

Logisim

Logisim is a very useful software tool that allows you to design simple as well as complex logic circuits. The architecture is highly modular, such that it allows you to add new modules without the need of recompiling the program. You can also add new modules even at runtime. Logisim can be used as an effective educational tool, as it calculates the logic flow through all gates recursively and redraws the screen at the end of all calculations, allowing for better understanding of any circuit simulation. *Supported OS: Windows, Linux and MacOS* **Link:** *http://sourceforge.net/ projects/circuit/files/2.7.x/2.7.1/logisim-win-2.7.1.exe/download*

Arduino IDE for windows

Arduino IDE for windows provides support for a wide array of your favourite Arduino boards, including Arduino Uno, Nano, Mega, Esplora, Ethernet, Fio, Pro or Pro Mini as well as LilyPad Arduino. The software works in conjunction with an Arduino controller and can be used to write, compile and upload code to the board. It is a modern alternative to other IDEs offering features such as syntax highlighting, automatic indentation and brace matching. It provides a streamlined interface with all of its features hosted inside a few buttons and easy-to-navigate menus, especially for professional programmers. *Supported OS: Windows* **Link:** *http:// www.softpedia.com/dyn-postdownload. php?p=104657&t=4&i=1*

KiCAD

KiCad is an open-source electronics design automation (EDA) package. It is a handy toolset for the design of electronic circuits schematics and their conversion to printed circuit boards (PCBs). With KiCad, you can create schematic diagrams and PCBs up to 16 layers. Apart from the included set of libraries, you can create your custom components as well. An integrated environment for all of the stages of the design process, namely, schematic capture, PCB layout, Gerber file generation/ visualisation and library editing along with 3D PCB viewing function makes this tool appropriate for any kind of project. *Supported OS: Windows* **Link:** *http://iut-tice. ujf-grenoble.fr/cao/KiCad_stable-2013.07.07-BZR4022_Win_full_version.exe*

ZamiaCAD

ZamiaCAD is a handy tool that provides a modular and extensible platform for your advanced hardware design, analysis and research. With its core components including a language-independent instantiation graph (IG) data structure, language-dependent frontend and the applications working on the IG data structure, it converts a hardware description (VHDL or Verilog) into a language-independent IG structure. The language-dependent frontend is extensible. *Supported OS: Windows, Linux and MacOS* **Link:** *http:// sourceforge.net/projects/zamiacad/files/ win32.win32.x86/zamiaCAD-0.11.3-win32_64.zip/download*

PVSim

PVSim is a portable and intuitive emulation utility for Verilog HDL. It features a fast compile simulate display cycle providing a fast simulation runtime with a user-friendly interface which allows you to open and simulate PSIM files and view the results within its main window along with all the simulation parameters and the event log. *Supported OS: Windows, Linux and MacOS* **Link:** *http://sourceforge.net/projects/pvsim/files/ PVSim/6.0.2/PVSim-6.0.2-win-setup. exe/download*

PCBColorizer

PCBColorizer is a small and useful tool that can be used to colourise your PCBs. This open source application can perform automatic scheme colouring of circuit boards. It works on the basic principle of marking each identical component with a unique colour and style, thus simplifying the visual search component on the board as well as the installation process. *Supported OS: Windows* **Link:** *http:// sourceforge.net/projects/pcbcolorizer/files/ PCBColorizer_1.07.zip/download*

ZenitPCB

ZenitPCB Layout is an excellent tool to create professional PCB. It is a flexible easy-to-use CAD program, which allows you to realise your projects in a short period of time. You can find an entire article on this software in this month's EFY Plus magazine. *Supported OS: Windows* **Link:** *http://www. zenitpcb.com/ZenitSuiteSetup180.zip* ●

The author is a technical journalist at EFY

Create PCBs up to 16 Layers with KiCad

Operational on Apple OS X, Linux and Windows, KiCad is an electronic design automation software tool to design and generate professional schematics and printed circuit boards. A well-known software tool, KiCad is released under the open source General Public Licence v2 (GNU GPL). Try out KiCad included in this month's DVD exclusively available with EFY Plus

ABHISHEK A. MUTHA

Designed and written by Jean-Pierre Charras, and under active development by the KiCad Developers Team, KiCad features an integrated environment that allows you to create schematic diagrams and PCBs up to 16 layers.

KiCad can be implemented and interoperated on multiple computer platforms. It is an open source software suite for EDA and it facilitates the design of schematics for electronic circuits and their conversion to PCB designs as well. Basically, it provides the user with an integrated environment for PCB layout design and schematic capture.

Tools exist within the package to create a bill of materials (BOM), artwork and Gerber files and 3D views of the PCB and its components too. A number of component libraries are available with the ability to add custom components by the users.

Interestingly, there are tools avail-

KiCad is considered mature and can be used for the successful development and maintenance of complex electronic boards. KiCad does not present any board-size limitation, and it can easily handle up to 16 copper layers and 12 technical layers. KiCad can also create all the files necessary for building printed boards like Gerber files for photo-plotters, drilling files, component location files and a lot more.

able to assist with importing components from other EDA tools like Eagle, for example. With multiple language support and 3D PCB viewing, there is a built-in basic autorouter as well.

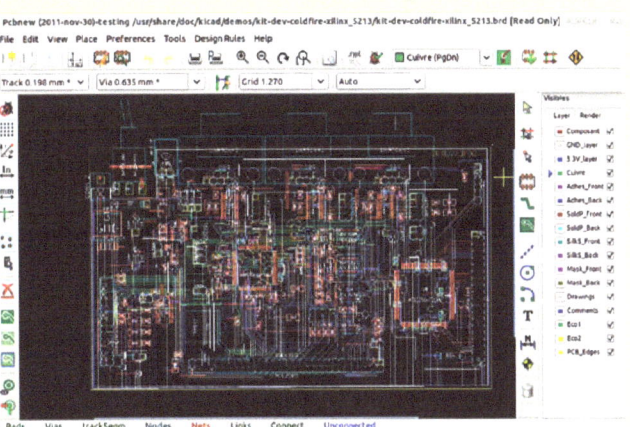

KiCad main window

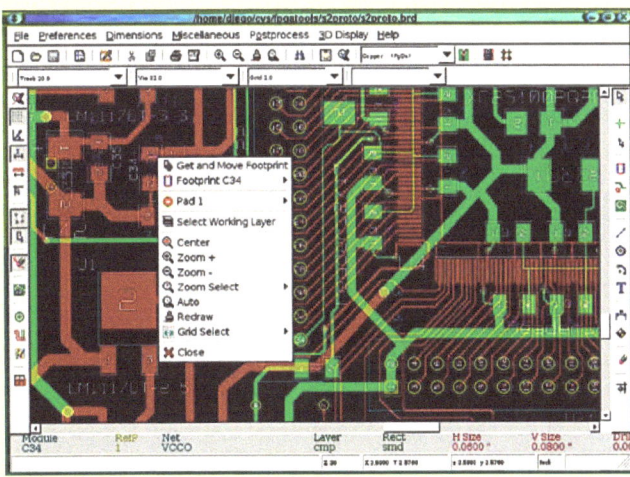

KiCad Pcbnew

Features and functionality in three steps

KiCad allows you to develop your printed circuit board (PCB) via three interconnected and independent main applications: Eeschema, Cvpcb, Pcbnew. This is simply done in three steps.

Step 1: Schematic capture. With the KiCad schematic editor Eeschema, you can create a sophisticated electronic sheet or a group of hierarchical sheets. Several schematic components come with the default KiCad library. An Electrical Rules Check (ERC) tool is available too.

Step 2: Component association. Cvpcb allows you to associate

KiCad overview

Beneath its singular surface, KiCad incorporates an elegant ensemble of the following standalone software tools. KiCad includes a project manager and comprises four main independent software tools:

Eeschema—the schematic capture editor.
Cvpcb—the footprint selector for components used in the circuit design.
Pcbnew—printed circuit board editor. It also has 3D view.
Gerbview—Gerber (photo plotter documents) file viewer.
Three utilities also included are:

Bitmap2Component. A component maker for logos (creates a schematic component or a footprint from a bitmap picture).

PcbCalculator. A calculator that is helpful to calculate components for regulators, track width versus current and transmission lines.

PlEditor. Page layout editor.

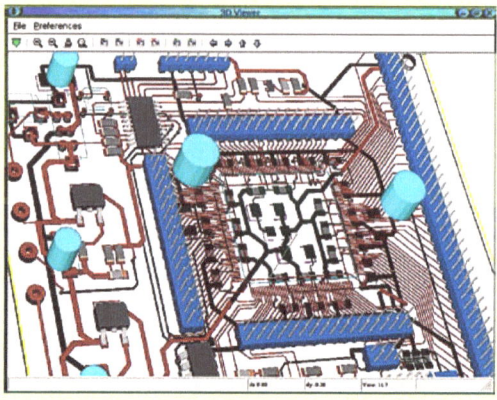

KiCad Pcbnew3D

Project

example.pro	project management file
example.sch	main schematic file
example.kicad_pcb	printed circuit board file
example.net	netlist file
example.xxx	various files created by the other utility programs
example.cache.lib	cache file of libraries used in the schematic (backup of the components used)

Being open source (GPL licenced), KiCad represents the ideal tool for projects oriented towards the creation of electronic hardware with an open source flavour.

each single schematic component with its footprint component. A very large component footprint library comes with KiCad.

Step 3: PCB layout. The Pcbnew board editor can handle up to 16 copper layers plus 12 technical layers (silk screen and solder mask) and allows you to layout the final PCB.

To successfully perform these steps,

KiCad comes with additional software tools. Two library component editors allow you to create or modify schematic components and footprint components. The 3D viewer allows you to render a final 3D model of your PCB.

For the generation of the necessary compliant files for manufacturing, your PCB (Gerber files for photo-plotters, drilling files and component location files), Pcbnew and Gerbview are used. Postscript or PDF file generation is also possible.

What one can do with KiCad

KiCad is considered mature and can be used for the successful development and maintenance of complex electronic boards. KiCad does not present any board-size limitation and it can easily handle up to 16 copper layers and 12 technical layers. KiCad can also create all the files necessary for building printed boards like Gerber files for photo-plotters, drilling files, component location files and a whole lot more.

Being open source (GPL licenced), KiCad represents the ideal tool for projects oriented towards the creation of electronic hardware with an open-source flavour. It is an excellent tool for professionals and hobbyist, mainly due to the excessive cost of commercial software that are prohibitive for students and hobbyists.

With the help of the project man-

ager, one can start the creation of a project by launching Eeschema. It manages a direct and fast access to component documentation. Eeschema is an integrated package because it comprises all the functionalities such as library management, layout, drawing and control to name a few. Using multi-sheet diagrams, it allows hierarchical drawings—simple, complex and flat hierarchies. Needed for modern schematic capture, Eeschema provides some essential additional functions such as:

1. Design rules check (DRC) for the automatic control of incorrect connections, and the inputs of components left unconnected.

2. Export of the layout files to POSTSCRIPT or HPGL format.

3. Printing the layout files on a local printer.

4. BOM generation.

5. Net list generation for PCB layout or simulation software.

KiCad component libraries

KiCad comes with a large set of open source library components. A text-based format is used for both schematic and PCB components. This allows the direct editing of your library files with any text-based software.

Both Eeschema and Pcbnew have a library manager as well as a library component editor for modifying and creating components and footprint parts. You can create, edit, delete or exchange library items easily. Documentation files can be associated to components and footprints, and key words, allowing a fast search by function. Very large libraries, created over many years, are available for schematic components and footprints. Most of printed board modules (footprint) are available with their 3D shape model.

The other very exciting aspect of KiCad is that library components for both PCB and schematic are actually plain text files.

KiCad's principle of use

In order to manage a KiCad project, schematic files, PCB files, supplemen-

About KiCad

Type: EDA
Initial Release: 1992
Latest stable release: BZR-4022 on July 7, 2013
Coded in: C++
Licence: GNU GPL v2, GNU LGPL v2.1
Operating systems: Linux, Microsoft Windows and Mac OS X (experimental)
Disk space required: 197MB (for Windows), 268MB (for UBUNTU)

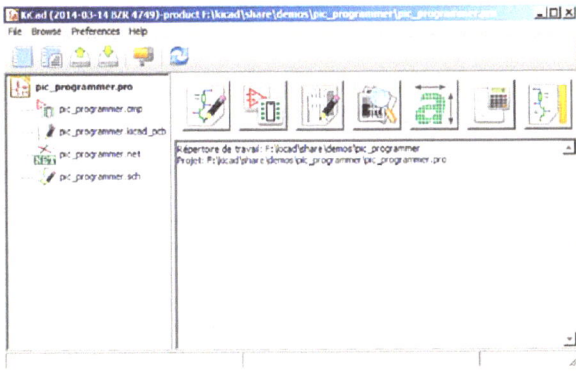

Fig. 1: Main window

Fig. 2: Utility launch pane

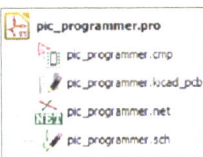

Fig. 3: Project tree view

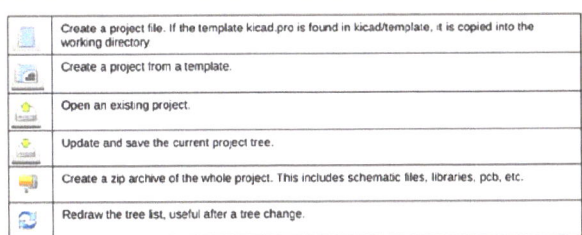

	Create a project file. If the template kicad.pro is found in kicad/template, it is copied into the working directory
	Create a project from a template.
	Open an existing project.
	Update and save the current project tree.
	Create a zip archive of the whole project. This includes schematic files, libraries, pcb, etc.
	Redraw the tree list, useful after a tree change.

Fig. 4: Top toolbar

and its directory. KiCad creates a file with a .pro extension that maintains a number of parameters for project management (such as the filename of the principal schematic, list of libraries used in the schematics and PCBs). Default names of both principal schematic and PCB files are derived from the name of the project. Thus, if a project called example.pro was created in a directory called example, the default files will be created as in the 'Project' box.

KiCad's graphical user interface

The main KiCad window (Fig. 1) is composed of a project tree view, a launch pane containing buttons used to run the various software tools and a message window. The menu and the toolbar can be used to create, read and save project files.

KiCad allows you to run all standalone software tools that come with it. The launch pane (Fig. 2) is made of the seven buttons that correspond to the following commands (from left to right): 1 – Eeschema, 2 – Cvpcb, 3 – Pcbnew, 4 – Gerbview, 5 – Bitmap2component, 6 – Pcb calculator, 7 – PI Editor.

Each project has a project tree view (see Fig. 3). Double-clicking on the

tary libraries, manufacturing files for phototracing, drilling and automatic component placement files, it is recommended to create a project as follows:

1. Create a working directory for the project (using KiCad or by other means).

2. In this directory, use KiCad to create a project file (file with extension .pro) via the 'Start a new project' icon.

It is strongly recommended to use the same name for both project files

Eeschema icons runs the schematic editor, which in this case will open the file 'pic_programmer.sch.' Double-clicking on the Pcbnew icon runs the layout editor, in this case opening the file 'pic_programmer.kicad_pcb.' Right clicking on any of the files in the project tree allows generic files manipulation. KiCad top toolbar (Fig. 4) allows for some basic files operation (from left to right).

Download and install KiCad

KiCad runs on Linux, Apple OS X and Windows. You can download a copy of KiCad from:

```
http://kicad.sourceforge.net/wiki/
Main_Page
```

Installation instructions are available on the KiCad website under: *Info/Install.*

Whatever installation method you choose, always go for a recent version of KiCad.

Under Linux. Under Linux, the easiest way to install KiCad is via Aptitude. Type into your terminal:

```
sudo add-apt-repository ppa:paxer/ppa
sudo aptitude update && sudo aptitude
safe-upgrade
sudo aptitude install KiCadKiCad-
doc-en
```

At the time of writing, the standard apt-get repository of Ubuntu offers a version of KiCad which is about one year old. Alternatively, you can download and install a pre-compile version of KiCad, or directly download the source code, compile and install KiCad.

Under Apple OS X. At the time of writing, the best way to install KiCad on Apple OS X was to download a pre-build binary from:

```
http://kicad.sourceforge.net/wiki/
Downloads
```

Note. Installation files for only Linux and Windows platforms are available in the DVD that accompanies this issue of Electronics For You Plus. ●

The author is a senior technical correspondent at EFY. The article has inputs from KiCad's latest documentation released on March 17, 2014

Create Professional PCB Designs with ZenitPCB Suite

A free, easy-to-use computer-aided design software package for Windows platform, that lets you design printed circuit boards of professional quality very quickly

ANAGHA P.

Several electronic design automation (EDA) software are available in the market, which provide professional-grade printed circuit board (PCB) designs. While the premium and full-featured versions of design software can be pretty expensive, there are several alternatives in the form of free integrated design environments (IDEs) available these days.

ZenitPCB is one such excellent EDA toolset that allows PCB designing for electronics projects. This freeware is flexible, has a simple user interface (UI) and lets you realise the design in a short period of time.

This design suite comprises four tools: ZenitPCB Layout, ZenitCapture Schematic, ZenitPCB Parts and ZenitPCB GerberView.

Why ZenitPCB

The ZenitPCB software package is absolutely free of cost and still provides designs with professional levels of accuracy. It comes with some advanced features offered by paid software — extensive and well-managed library, multiple layers and double-sided designs support, high flexibility, superior schematic design, compatibility with standard design file formats and a tool for viewing Gerber file.

The software reduces the time to market; for example, the user can import a mechanical drawing with a PCB and build his/her board outline, thereby avoiding mistakes and saving time.

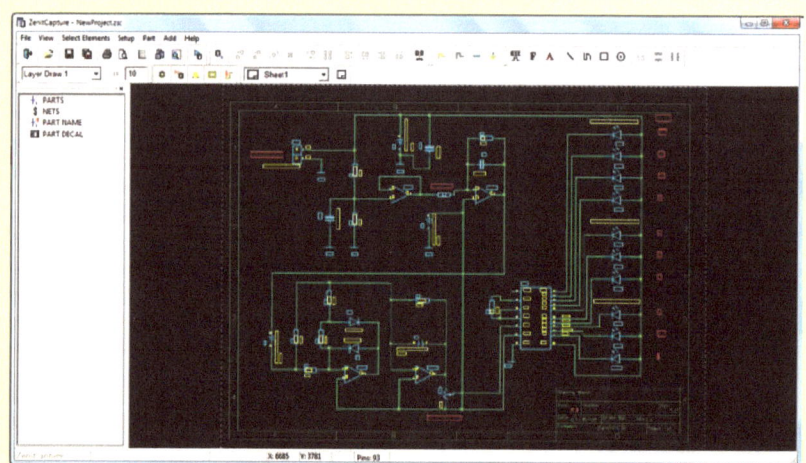

Fig. 1: Screenshot of ZenitCapture project

Who can use it

ZenitPCB Suite and the accompanying libraries are licenced as freeware and can be used for personal, educational and semi-professional purposes. It is aimed at designers, engineers, hobbyists, students and academicians who want to create their own PCBs with a professional touch, without having to pay for expensive licences. The circuit board for design has 800 pins and serves the level of complexity that would be required by designs for the targeted audience. This 800 pins limit, according to the developer, is the borderline between the hobby projects and professional jobs.

Designing schematics with ZenitCapture

Schematic designs can be created quickly and easily using the feature-rich toolset of ZenitCapture.

Schematic is a diagram that acts like a map for designing, building and troubleshooting electronic or electrical circuits. It gives us the plan of components used for a design (represented as electronic symbols) and how they are connected.

A schematic in ZenitCapture can have several layers and the layout is created in a netlist file format. The part libraries have a collection of more than 800 commonly-used components (both surface-mount devices and through holes) suitable for immediate placement on the project. The device required for design can be either selected from these libraries or edited/created using ZenitParts.

Connections can be easily added with orthogonal locking and snap-to-pin features. You can also place global symbols (ground pin – GND and power supply pin – Vcc), IEEE symbols

Customising the ASCII report

The ASCII report can be customised based on the following:

Bill of materials. The ASCII report will be sorted based on factors such as component name, type of materials, consumables (for example, solder and solder paste) and routing information. Useful for determining the amount of each component required for the designing.

Part list. This list is for browsing only the top-level components and avoids material, routing information and consumables.

Component location. Gives you details of the position of each component—its position and angle in X-Y axes, the side of PCB on which they are placed (top or bottom).

PickPlace location (ppl). This file is required to automatically assemble the PCB according to the X-Y coordinates and relative positions.

Jumpers info (txt). Used to find out the jumpers used in the design.

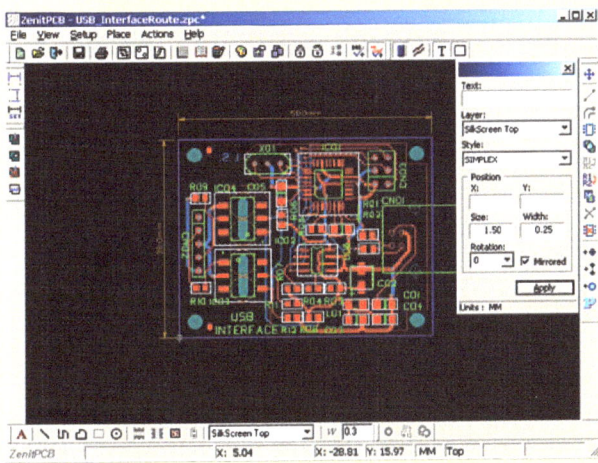

Fig. 2: A complete PCB design using ZenitPCB

(Sigma – Σ, Pi – Π, bidirectional arrows – ↔, hysteresis – ∪ etc), off-page connector symbols, wires, shapes, texts, field texts and so on in the schematic. Other features include auto-renaming of reference, ability to mirror the parts, customisable colour schemes and visibility for each of the layers, availability of datasheets on parts provided in the library, option to export netlist format files into PCB and customisable ASCII report.

ZenitPCB allows the export of reports into ASCII files into either no separation (as image) or separated (comma-separated values or CSV—to import into spreadsheet).

ZenitPCB layout tool lets you use Netlists from competing software

This tool allows the user to import a netlist file from various other major schematic capture software (OrCAD by Cadence Design Systems, Protel/Altium

Designer by Altium Limited, EAGLE by CadSoft Computer, NI Multisim by National Instruments Electronics Workbench Group, PADS by Mentor Graphics, to name a few) while preserving all the components and their relative electrical connections.

Design without schematic. It is also possible to create the PCB design project directly from the layout editor, without the need for importing netlist from schematic capture. This can be done by importing components from the library and then connecting those pins using the cursor (auto ratlines).

Jumpers. Jumpers are short pieces of conductors that let you bypass part of a circuit and make electric connection between two pin headers. The current version has an autojumper introduced to it. When routing, you can import jumper decals into the project and place the selected surface-mount jumper on the top or bottom side of the PCB.

Photo View. This view gives you the idea of how the real PCB would look like; it shows all the through holes, slots and silk screens. The user can check whether the components are placed in the correct side, the silk and references are placed well or end up on the pads.

Netlist format is basically an ASCII (American Standard Code for Information Interchange) text file that contains electrical and electronics information.

Signal trace is the copper foil that remains after etching a PCB. It conducts signals and is equivalent to connection wires.

Rename Components. The AutoRename Reference command allows you to rename all the components in the layout. You could choose whether they should be numbered horizontally or vertically, and the width of the band. Also, you will be able to create ASCII for back annotations of several schematic captures.

Import/Export DXF. It is possible to import and export every layer and associated items from and to DXF files compatible with commercial CAD such as AutoCAD, IntelliCAD and TurboCAD.

Export IDF. It is possible to export the design created in ZenithPCB to ASCII file containing all the 3D information. This file is compatible with most professional CAD software, making it useful for mechanical designers.

Trace Current Capacity. The Current Capacity option lets you know how many amperes of current a trace could carry, by choosing the empirical formula, foil thickness and temperature rise above ambient condition.

Alternative Bottom Component. This field in component library lets you set the bottom footprint for both wave soldering and reflow soldering.

Obstruct Shapes. It is possible to define areas in the PCB where certain operations such as copper pour, placement and routing are not allowed. Height obstruct option is used for checking against component keep-out areas.

Padstack. New features allow the user to make pad in an offset shape, and with slot holes.

You can make both single-sided and double-sided PCBs using ZenitPCB Layout. The component library has more than thousand footprints. The footprint wizard makes the creating of

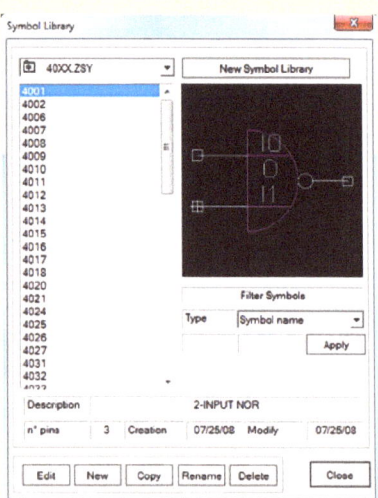

Fig. 3: Symbol library for ZenitParts

footprint easier. Design rule checker considers the rules to be followed while placing elements in a PCB, like the clearance between board outline, tracks, pads, components and text.

ZenitParts

ZenitParts is a simple tool that allows you to quickly create and modify parts and symbols for PCB designing purposes. The user-friendly and intuitive UI, very similar to that of ZenitCapture schematic, makes this program easy to understand and use. The created parts and symbols can be easily saved to the library of our choice. It also allows you to modify the items already provided in the library, copy, delete and rename them.

ZenitPCBGerberView

The default file format *.zpc created in ZenitPCB Layout can be exported to Gerber and NC Drill file formats from

Gerber is an open, 2D vector image file format used by PCB designing software used to store and communicate design information such as copper layers, drill holes and solder mask.

The Excellon NC Drill is a file format that is used in PCB drilling and routing purposes, and defines machine-specific information such as the drill feed and speed.

File > Export and selecting the option required. This opens the *Generate Gerber Files* or *NC Drill Format* dialogue box accordingly. Along with the Gerber and NC Drill files, the system creates an ASCII file with the name and folder of the Gerber or NC Drill files. Once this is done, click on the *Open GerberView* icon from below the menu bar to open the *ZenitPCB GerberView* window, and it displays the Gerber format according to the ASCII file.

The GerberView has options to hide and view separate layers or all at once. This makes analysing and editing easier for the design. The colour view of each layer can be configured according to the user's preferences. When *Overlay Display Color* is activated, it can control the transparency of various layers so that the user can observe more layers at the same time.

Steps for creating PCB design

Here are the basic steps that need to be followed when you are designing a PCN using ZenitPCB suite: First place the components in ZenitCapture. Route all the components. Package the parts and create a netlist. Now create a new project or open an existing one in

Licence type	Freeware
Developer	StortiniMirko Bruno, www.zenitpcb.com
Latest version	1.8
Operating system	Windows (all versions)
Disk space	33.8MB (excluding resources required to save the results)

ZenitPCB Layout, place a board outline (*Place > Board Outline > Wizard… / Polyline*). Import the netlist from *File > Import > Netlist…* option. Now you can route the circuit to create a suitable PCB design.

On an alternate method, you can directly design without creating schematic, by importing and/or editing components from library using File menu.

Easy user interface and detailed help files

The basic steps for designing using ZenitPCB suite are very simple and require little documentation. But if you had selected to install help file along with the package, then you can get access to an elaborate help file that guides you to create a PCB design right from the beginning. The help window can be opened by pressing the function key F1. The download page of the developer site www. zenitpcb.com also has a collection of video tutorials. Both these would help a novice engineer to get through the design process without much trouble. For more tips, advices and guidance, you could send a mail to support@ zenitpcb.com. ●

The author is a technical correspondent at EFY

ZamiaCAD, an Open Source Platform for Advanced Hardware Design

ZamiaCAD is a modular and extensible platform for your advanced hardware design, analysis and research. Let us take a look at what it delivers

PANKAJ V.

If you are looking for an open source application for real-world hardware design, debug and analysis, ZamiaCAD could be the right solution for you. ZamiaCAD has been implemented with high performance, scalability and usability as the main targets, and it has already become a basis for VHDL IDE Eclipse plug-in with built-in simulator.

Significant advances have been made in software development discipline in the area of program code entry, navigation and analysis, while the existing hardware development environments are still lacking this progress, especially in the open source tools domain. ZamiaCAD is an open source platform that aims at transferring the advances in software development to the hardware development discipline.

How it works

This multi-platform tool provides a modular and extensible framework that can also be used for design entry, navigation and analysis along with the simulation functions. Besides, the cache and persistent data structures make this platform capable of handling very large industrial designs.

If we look at the structure of ZamiaCAD, it can be broken into various building blocks, such as frontend, its core and an Eclipse IDE plug-in based GUI, with various features providing a handy environment for the

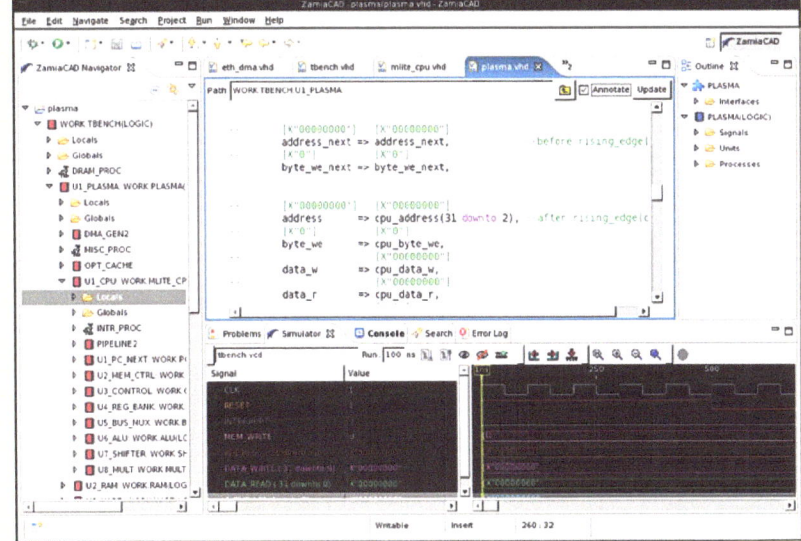

ZamiaCAD

advanced hardware designing.

The language-dependent modular and extensible frontend for HDL parsing consists of a parser and an elaboration engine, which help design engineers utilise the advances in software development in the hardware development discipline.

Currently, full VHDL elaboration is included with a complete frontend.

But Verilog has only a parser with persistent and scalable syntax tree storage backing both these frontends, which helps these language-dependent frontends to generate language-independent instantiation graphs (IGs), accounting for the scalability of intermediate design representations in terms of size and performance.

Responsible for the intermediate

An interpreter for elaboration

Computation of various static values, like the range of boundaries for the system type or conditions and the loop ranges at the time of unrolling generate statements, is done at the time of elaboration. For easy and fast computations, ZamiaCAD includes a complete interpreter implemented as a stack machine supporting all operations that can occur during elaboration, which include math and logic operators and even (potentially recursive) subprogram calls. This quick expression-evaluation interpreter is implemented in the package *com.⬚amaicad.inst⬚raph.interpreter.*

design representation and analysis, the core is based on a powerful, persistent and scalable design database, ZDB. Unlike other open source tools, ZamiaCAD has its own built-on object database based on Java serialisation technology, helping design engineers with its fully elaborated design model and the useful static analysis support tasks, such as global signal reference search and FSM recognition. Additionally, the core also includes some waveform file import modules and a built-in simulator for ad-hoc validations.

The simulator applications, along with the Eclipse GUI, are built on top of the IG and the language-dependent abstract syntax tree structures in the frontend. This supports graphical viewers and editors with an automatic model builder for their ease of modelling and design by utilising all the functionality of the other blocks.

Another significant feature of ZamiaCAD comes in the form of a very high-speed VHDL indexer that extracts the information about which design units are declared in which files. The actual VHDL parser can then be used whenever required on local source files, or those in your eclipse workspace, and the files defining design units that are actually required during elaboration.

Features for your ease of use

ZamiaCAD is a tool for advanced hardware design, analysis and research. It can be beneficial for developers writing EDA tools, end-users using EDA tools coming with the platform and also the researchers prototyping research algorithms. ZamiaCAD supports many

Special advantages

ZamiaCAD adheres to its high performance, scalability and usability goals by its various design principles. It features many advantages over its competitors.

Non-invasive designing. One of the major design principles of ZamiaCAD is non-invasiveness. Any engineer can start working with ZamiaCAD using pre-existing projects with minimum configuration effort as HDL being one and only data format. It also ensures that no other member of the team is forced to work on ZamiaCAD if one is using it.

Scalability and high performance. Scalability and high performance targets are ensured by the IG structures used for intermediate design representation along with a powerful, persistent and scalable design database.

Supports large designs. ZamiaCAD can also transcend memory limitations in case of particularly large designs as it is empowered by efficient Java serialisation technology.

Robustness. Another advantage of Java is that, JUnit tests exist for many parts of the Java framework, which accounts for robustness of the tool.

features that can be of great help while working with the tool, such as:

Full design-entry support. ZamiaCAD supports syntax highlighting and auto-completion functions with the help of easily-extensible VHDL templates and also the outlines. These design-entry-support functions account for an incremental model builder approach and easy error marking.

Easy design navigation. ZamiaCAD allows you to navigate easily through your designs and trace your signals and references. You can open your design units and signal declarations using instantiation graph (IG) and hierarchical path.

Handy simulator. The built-in reference simulator, along with waveform viewer and waveform file import functionality, accounts for handy simulations in ZamiaCAD. You can also have source back-annotation of simulated values.

Complete design analysis. Design analysis can be seen in parallel with navigation. You can have reference search for instantiated modules and signals across hierarchy levels in the design. It covers almost all the features of

any Eclipse framework with additional functionalities like declaration search and file caching to mitigate latencies.

A powerful non-invasive tool

ZamiaCAD is a powerful non-invasive tool made for design and verification engineers. It is well suited for the automation of manual design and verification tasks in the design areas like entry, analysis, integration of designs and simulations. This can help in increasing register transfer level design and verification engineers' productivity.

With the complete VHDL standard supported as hardware description language and options to extend it by adding other frontends, such as Verilog, ZamiaCAD qualifies well for being a good option for advanced hardware designs.

With its structure and design principles, ZamiaCAD tool supports the state-of-the-art and future large-scale industrial hardware designs. It is equipped with powerful debugging functionality and provides the research community with an open source robust research platform.

Support

ZamiaCAD is a free GPL licensed software tool that can be downloaded at http://zamiacad.sourceforge.net/web/. Also, you can find developer's documentation and some quick-start tutorials available on the website for you to get started with ZamiaCAD and know how it looks and feels like. ●

Installing ZamiaCAD

Installation file can be found in EFY Plus DVD that came with this magazine. The DVD includes ZamiaCAD bundled as a product that includes Eclipse with embedded ZamiaCAD plugin.

Supported OS: Windows, Linux, MacOS

HDD space required: About 100MB

Note. If you already have Eclipse running and you want to integrate ZamiaCAD into your Eclipse, you can install ZamiaCAD plug-in using the following steps: Open menu Help→ Install New Software... and enter location for your ZamiaCAD files, for both Name and Location in the Add window. Alternatively, you can also enter the website url of the latest version of ZamiaCAD download location. After completing installation, open Window→Open Perspective→Others→ZamiaCAD.

The author is a technical journalist at EFY

SPEED   SECURITY   SELF RELIANCE

Transforming the Value Chain... Realising Opportunities

ORGANISED BY

ELCINA ⊛

ELCINA Electronic Industries
Association of India

EVENT HIGHLIGHTS

- Two days Exhibition
- Conference
- Buyer - Seller Meet
- Special Business Promotion Sessions

FOCUS SEGMENTS

Military - Land Systems
Aerospace - Air force, Avionics
Naval Systems
Homeland Security

"India is set to undertake one of the largest equipment procurement cycles in the world with an estimated spend of about USD 112 billion on capital acquisitions by the year 2016, which will create offset opportunities for the domestic industry worth USD 30 billion"*

(*source KPMG Report)

30th - 31st July, 2014
DEFENCE & AEROSPACE
SES 2014
5th Strategic Electronics Summit
Bangalore International Exhibition Center,(BIEC), Bengaluru , INDIA

Conference Theme - "Make Indian- Dream to Reality"
Transforming the Indian Strategic Electronics Ecosystem

DAY - I

The Road to Indigenization

- Role of Policies in Indigenization – DPP
- Challenges of Technology Transfer for Indigenization
    - Gaps in Technology & Potential Sources

Regulatory Issues

- Offsets - Implementation & Regulation
- Challenges faced by SME's

DAY - II

Defence & Aerospace Market Potential

- Key Defence Procurement Programs with High Electronics Content
- Defence Offsets and their Business Potential
- Capabilities of Indian Industry in Defence Electronics

Silver Sponsor

सी-डॉट
C-DOT

Proposed Report on
"Systematic Analysis of Electronics Content in Various Defence Programs"

Media Partner    Supporting Media    Segment Partner Homeland Security    Supporting Association IACC INDO-AMERICAN CHAMBER OF COMMERCE   CEO Networking Dinner Sponsor

Registration: To Register as a Delegate and / Or book a stall please contact:-

Rajesh Rawat: Mobile: +91 9911445890 Tel: +91 011 41615985 Email: rajesh@elcina.com Fax: +91 011 26929440

ELCINA Electronic Industries Association of India

ELCINA House, 422 Okhla Ind. Estate, New Delhi, INDIA - 110020 URL *www.elcina.com or www.sourceindia-electronics.com*

How to Select AA-Size Battery Cells

Electronic devices run on batteries of different shapes and sizes. But the most popular amongst them are AA-size primary cylindrical cells. These cells' prices are different but for most buyers they all are the same. But does a cell with higher price last longer? To find out we did a simple accelerated test

VIVEK PANCHABHAIYA

Modern battery cells use a variety of chemicals to power their reactions. Common battery chemistries include:

Zinc-carbon. These are the lowest-cost primary (non-rechargeable) cells meant for household use. The anode is zinc, the cathode is manganese dioxide and the electrolyte is ammonium chloride or zinc chloride. These cells produce very low power, but have a good shelf life and are well suited for clocks and remote controls.

Alkaline. These are the most commonly used primary cells. The cathode is made of manganese dioxide, while the anode is zinc powder. These cells have derived their name from the potassium hydroxide electrolyte, which is an alkaline substance. These provide more power-per-use than zinc-carbon and secondary (rechargeable) battery cells, and have an excellent shelf life.

Lithium. These are primary cells with a lithium metal or a lithium compound as an anode. These offer performance advantages well beyond the capabilities of conventional aqueous electrolyte battery systems. Their shelf life can be well above 10 years and they work at very low temperatures. These are mainly used in small formats (coin cells up to AA size) because bigger sizes of lithium batteries are a safety concern in consumer applications and are only used in military applications.

Silver oxide. These are also primary batteries with relatively very high energy/weight ratio. Their cost is linked to the price of silver. These are avail-

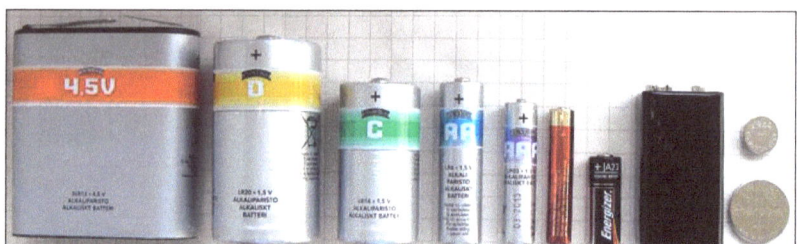

Fig. 1: Different sizes of batteries

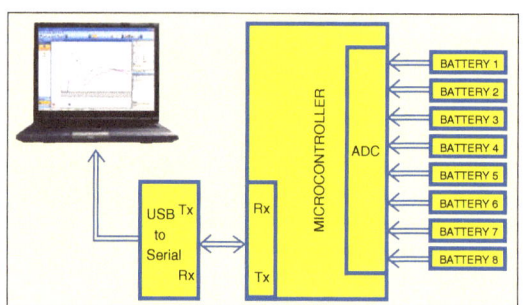

Fig. 2: Block diagram of the test setup

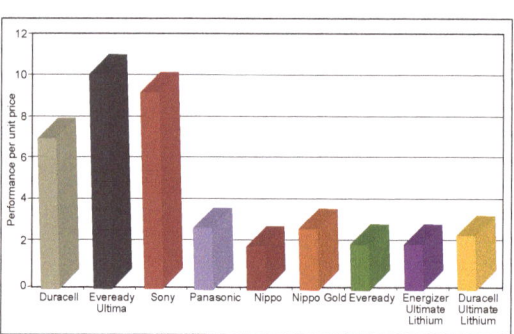

Fig. 3: Performance per unit price

able in either very small sizes as button cells, where the amount of silver used is small and not a significant contributor to the overall product cost, or in large custom-design batteries, where the superior performance characteristics of the silver oxide chemistry outweigh cost considerations.

Nickel-cadmium. These are rechargeable or secondary batteries

which use nickel oxide hydroxide and metallic cadmium as electrodes. The abbreviation Ni-Cd is derived from the chemical symbols of nickel (Ni) and cadmium (Cd). Ni-Cd batteries are rugged and reliable. These exhibit high-power capability, wide operating temperature range and long lifecycle, but have low runtime per charge.

Nickel-metal hydride. A nickel–metal hydride battery cell, abbreviated as NiMH or Ni–MH, is also a rechargeable cell. It uses positive electrodes of nickel oxyhydroxide (NiOOH), such as the NiCd, but the negative electrodes use a

TABLE I
Types of AA-Size Battery Cells

Types	IEC name	Nominal voltage	Rechargeable
Zinc-carbon	R6	1.5V	No
Alkaline	LR6	1.5V	No
Li-FeS2	FR6	1.5V	No
NiCd	KR6	1.2V	Yes
NiMh	HR6	1.2V	Yes
NiZn	ZR6	1.65V	Yes

TABLE II
Test Results

IEC name	LR6 (Alkaline)			R6 (Zinc-carbon)				FR6 (Li-FeS$_2$)	
	Duracell Copper Top	Sony Stamina Plus	Eveready Ultima	Panasonic	Nippo Gold	Eveready	Nippo	Energizer Ultimate Lithium	Duracell Ultra Lithium
Model number	MN1500	AM3	2115	UM-3NG	UM-3DG	1015	UM-3U	L91	LF1500
Price (₹)	28	20	20	10	10	11	7	175	170
Sustained duration (in hours)	3.8258	3.4108	3.8944	0.8736	0.8533	0.7144	0.4253	7.24	8
Load					2.5Ω				
Minimum predefined voltage					0.8V				

hydrogen-absorbing alloy instead of cadmium, being in essence a practical application of nickel–hydrogen battery chemistry. An NiMH cell can have two to three times the capacity of an equivalent-sized NiCd, and its energy density approaches that of a lithium-ion cell.

Lithium ion. A lithium-ion cell (sometimes called Li-ion battery) is a rechargeable battery cell in which lithium ions move from the negative electrodes to the positive electrodes during discharge and back when charging. These batteries use an intercalated lithium compound as the electrode material, compared to the metallic lithium used in non-rechargeable lithium batteries.

Lead-acid. These are the most popular rechargeable battery cells worldwide. Both the battery product and the manufacturing process are proven, economical and reliable. However, as these batteries are heavy and use a liquid electrolyte, they are not used in portable consumer electronics.

Technical standards for battery sizes and types are published by standards organisations such as International Electrotechnical Commission (IEC) and American National Standards Institute (ANSI). Some sizes of batteries are D, C, AA, AAA, AAAA, A23, 9V and CR2032.

Accelerated testing

Different AA-size primary battery cells were chosen for testing in EFY lab. These cells come with a variety of chemistries and therefore the difference in their prices. You can identify

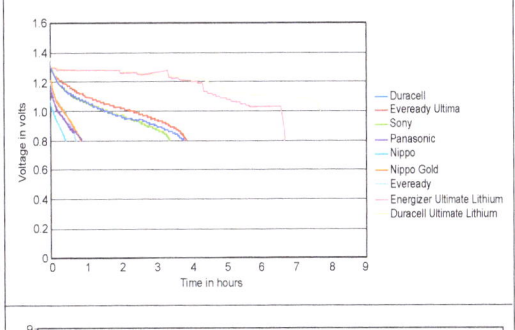

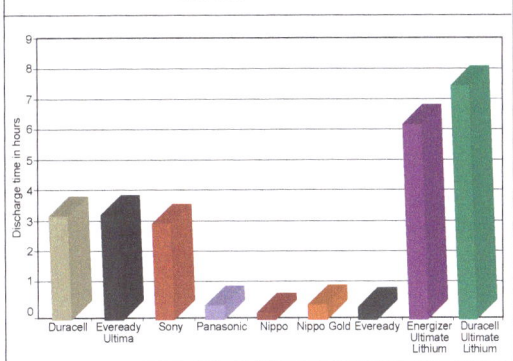

Fig. 4: Real-time data (up to 0.8V)

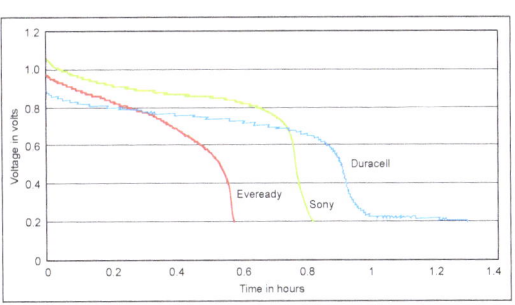

Fig. 5: Real-time data (0.8V to 0.2V)

different cells with their IEC names as shown in Table I.

Fig. 2 shows the test setup for accelerated testing of various AA-size primary batteries to have a rough idea about their performance vis a vis their price. Our testing was simple. Different batteries were put on

the same load (2.5 ohms). The voltage of each battery was continuously fetched by a microcontroller (PIC16F877A). The microcontroller sent the real-time voltage levels of all the battery cells to a MATLAB program running on a computer via serial port. The program saved these values in an Excel sheet for all the cells. The program stopped the process when all the batteries' voltage fell below the minimum predefined level. The program provided the result in number of hours it took to discharge up to that level. Here 0.8V was taken as the minimum predefined value.

Table II shows the test results for different makes and chemistries. The data shows that Eveready Ultima (LR6) gave the best performance per unit price as shown in Fig. 3.

Fig. 4 shows the real-time graphs generated by the software. But when the test was run further in the range of 0.8V to 0.2V, the Duracell sustained much longer amongst all alkaline (LR6) batteries. Other batteries degraded very quickly after 0.8V but Duracell sustained much longer as shown in Fig. 5. ●

The author is a □.Tech in electronics and commu□ nications from □□CEM Colle□e, □□alior

Electronics Manufacturing Jobs to Multiply by 2020

Here we explore the kind of job opportunities available in electronics industry in India, specially in the surface mount technology (SMT) sector. Also covered are skills required and pay packages that may be expected

ABHISHEK A. MUTHA

Electronics manufacturing industry in India needs to grow at a much faster pace due to the ever-growing demand for consumer electronics, IT and telecom goods. This demand is expected to grow to US$ 400 billion by 2020, according to some estimates. In case manufacturing in India does not keep pace with the demand, country's import bill for electronics will exceed the import bill for crude oil.

Amarpreet Singh, manager–manufacturing, Centum Electronics says, "To support this demand and reduce the currency outflow from India, Indian government has come up with a very detailed industrial policy that encourages manufacturing industry in India. Indian manufacturing industry is growing and so are the skill requirements to support electronics industry."

Four broad fields for manufacturing

In electronics manufacturing value chain, there are four broad fields where electronics engineers fit in, namely, electronics tool development software industry, design organisations (who use these tools to develop electronics designs), original equipment manufacturers (OEMs, who use the designs and integrate them in their product systems themselves or through electronics manufacturing services providers) and, finally, the qualification and failure analysis labs, according to Ankan Mitra, vice president, SMTA India Chapter.

Software tools industry for electronics design and manufacturing, and second, components, PCBs and system design houses are the two major fields where opportunities for engineers exist at present. Mitra says, "There is a huge industry for people with an electronics background, but it is aligned towards software tools development for electronics design industry. There is also a huge requirement for talented engineers in design companies."

Next come the OEMs, who use services of all these design companies to manufacture their products in their own factories or outsource it to electronic manufacturing services (EMS) companies. The fourth broad field to make a career in is failure analysis and qualification labs in India. Mitra informs, "In case of field failures, specialist engineers come into picture. Field engineers are specialised engineers who identify issues and report problems faced by customers and service engineers. They are in-charge of carrying out critical repairs on faulty machines and they play an important role in the electronics manufacturing field as well."

Apart from these, there are component suppliers or traders who exist across this value chain. Mitra says, "Engineers procure components during the NPI cycle, manufacturing cycle and even during field failures. These are the five different layers, with the component supplier organisations being a parallel layer, which constitute different segments in the electronics manufacturing industry. Each segment provides different opportunities for engineers."

From design, production, test to SMT

The opportunities in electronics manufacturing industry are immense for fresh engineers as well as experienced professionals. "Entry-level roles vary widely depending on the company, products and geographies. But, usually, it involves assisting a team of mid-level managers who work on product development and manufacturing," informs Kaustubh Nande, manager–marketing, ANSYS Software (ANSYS India).

Sharing his views on opportunities in this field, T. Anand, principal consultant, Knewron says, "Opportunities exist in design, testing and assembly domain for engineers interested in these fields. However, entry-level roles in this domain focus on hands-on work rather than pure design or coding assignments. This means, one must understand and appreciate the importance of practical work experience before progressing in this field and making long-term career."

Surface mount technology (SMT) provides a wide and diverse field in electronics industry, either directly or through ancillary producers, explains Vivek Madhukar, COO of Times Business Solutions, which operates TimesJobs.com. He says, "The biggest employers are in IT, manufacturing/engineering, healthcare/biotechnology and auto sectors based in and around Bengaluru, Pune and Delhi. Salaries start at the lower end, but hands-on experience definitely improves one's growth prospects, with earnings reaching over 1.5 million rupees with adequate exposure."

Amit Bhargava, managing director, Asha Electronics believes the key to boost manufacturing sector lies with fresh graduates. He says, "Engineers need to realise that the real fun is in creating, that is, manufacturing. Young graduates can fire up the industry with their energy and ideas. Entering the realms of purchase, production, quality and marketing will energise these streams." He adds, "As production gets complex so will marketing, and this will help young graduates pick up lucrative assignments and incentives from companies. Starting as trainees, and setting themselves up in fields they excel in, will boost their growth prospects."

"Most of our designs are SMT-based and need high skill levels in assembly. Being a design house, we do hand-soldering as much as possible and depend highly on practical experience," says Anand. He adds, "At Knewron we welcome engineers from all strata; however, each one needs to demonstrate his worth before coming on board." Talking about his company, Bhargava informs, "Engineers are as such encouraged in our organisation to take up jobs in all functions. The field being complex, engineers are best suited for it."

Centum Electronics recruits as well as encourages fresher engineers to work on their state-of-the-art SMT lines. Srinivasa Bharadwaj, manager-engineering, Centum Electronics says, "As the SMT machines involve understanding of electronics assembly processes as well as maintenance of the pick-and-place equipment, where a lot of mechanical movement of the parts is involved, we prefer to have engineers from electronics and mechanical streams. Engineers from electrical stream also fit in."

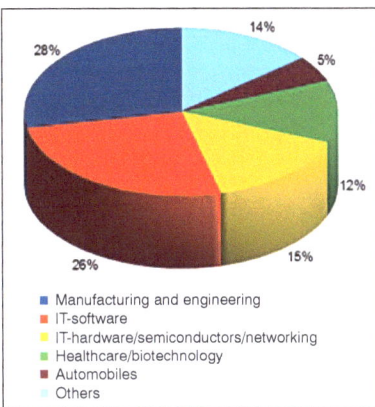

Fig. 1: Top industries hiring professionals for electronics manufacaturing

- Manufacturing and engineering
- IT-software
- IT-hardware/semiconductors/networking
- Healthcare/biotechnology
- Automobiles
- Others

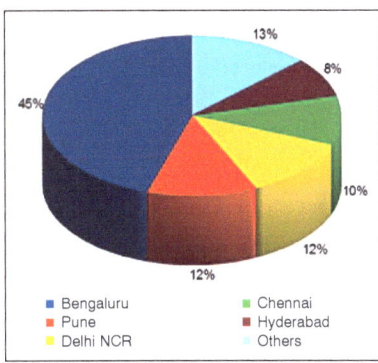

Fig. 2: Top locations for jobs in electronics manufacturing

- Bengaluru
- Pune
- Delhi NCR
- Chennai
- Hyderabad
- Others

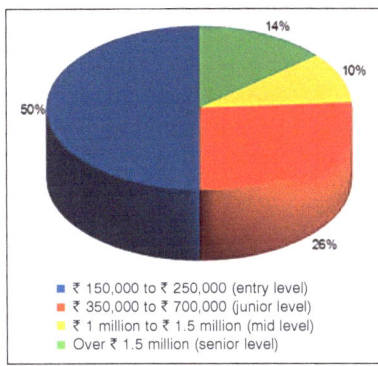

Fig. 3: Salary break-up for engineers in electronics manufacturing segment

- ₹ 150,000 to ₹ 250,000 (entry level)
- ₹ 350,000 to ₹ 700,000 (junior level)
- ₹ 1 million to ₹ 1.5 million (mid level)
- Over ₹ 1.5 million (senior level)

Why a career in SMT

Fresh engineers interested in learning equipment functionality and process engineering are suitable for SMT lines. In addition, engineers with an aptitude of equipment maintenance can also take up the job of an SMT engineer. Singh says, "As it is the core of an electronics manufacturing industry, there is a lot of focus on the technological developments in this area. This makes it more interesting

and challenging for any of the fresh engineers to take up SMT as a career option."

According to Singh, SMT forms the back bone of electronics assembly line and is therefore a very important area for an electronics manufacturing plant. He says, "Though there is a requirement of experienced engineering resources running SMT assembly lines, it is very important to have a significant pool of fresh engineers working on SMT lines too." To that Bharadwaj adds, "Engineering graduates with a good aptitude for learning basic manufacturing processes like screen printing, solder-paste inspection, pick and place, reflow soldering and automatic inspections are preferred. It is very challenging for a fresh electronics engineer to work on fully-automatic computer-controlled equipment. The industry offers various roles, which include responsibilities like process engineering and equipment maintenance of the SMT lines."

What an SMT engineer learns

Talking about the exposure for an SMT engineer, Singh shares, "Since SMT assembly is the core of electronic assembly, an SMT engineer gets exposed to critical process areas like tooling design (such as stencil, palettes), pick-and-place programming, reflow profiling, AOI programming and x-ray inspection." Shifting to the equipment side, he adds, "Here, the areas where an engineer would get an insight into are breakdown maintenance, equipment effectiveness, equipment capabilities and new equipment selection, to name a few."

Learning about statistical process controls for critical processes like solder-paste printing, for instance, using very-high-end inspection equipment is another interesting area in SMT. "SMT first-pass yield is critical in electronic assembly. Hence, invariably, the SMT engineers have to learn about the process to bring in yield improvements. Therefore there is opportunity to learn and use statistical techniques to bring forth yield improvements," informs Bharadwaj.

Aptitude, right attitude and soft skills

According to Anand, for SMT-assembly and manufacturing-based jobs, people must have practical experience in soldering. He informs, "Even a fresher should not only have worked during his college assignments but even otherwise have practiced enough to demonstrate that he is interested in this domain. Assembly, troubleshooting and testing are some of the widely sought skills for this domain."

A good understanding of basic electronics is required. Mitra says, "It's not just the engineering degree, a potential candidate needs to understand how a design looks like and how to interpret the design so that during manufacturing, in case there are some issues, these can be reported. That's the kind of value add that people look for in profiles of the engineers." He adds, "There are also multiple industry certifications that recruiters tend to look at. There are some certifications that engineers might have acquired in the past which could help them stand out among other candidates."

From a product-development perspective, physics and engineering tool knowledge are extremely important, informs Nande. He says, "If you are on the manufacturing side, obviously process and supply-chain knowledge is useful. Increasingly, soft skills and good communication skills are also seen as differentiators in promoting engineers to managerial cadre."

Demand areas

According to the data provided by TimesJobs.com, the top industries hiring engineers are in the manufacturing and engineering sector with about 28 per cent share of jobs, followed by the IT-software/software services sector with a share of 26 per cent. IT-hardware/semiconductor and healthcare/biotechnology sectors provide 15 per cent and 12 per cent of the jobs, respectively.

Apart from Bengaluru that features in the list of top locations for jobs in electronics manufacturing and SMT sector with 45 per cent share, figures indicate Pune, Delhi and Chennai provide 12 per cent, 12 per cent and 10 per cent of the jobs, respectively. Remaining jobs are scattered in other locations.

Pay package

This industry will pay well, but only if you have patience. As in any other manufacturing industry, here too, you must be ready to toil, informs Bhargava. Creation of product

Are certifications worth it?

Yes, certifications are worth it, but mostly at a slightly later stage of career. Normally, these certifications are quite expensive, and though most students acquire these as a part of their engineering curriculum, usually, it is not a viable proposition, according to Mitra. He says, "If somebody is studying and thinking of doing a certification, it is very difficult for recruiters to interpret what the certification is intended to do. Normally, what is preferable is to gather a solid experience of about one or two years and then only go for certification courses. These certification courses cost a significant amount of money in India. Also, it would be difficult for students to interpret, along with their studies, what the trainer is trying to teach."

does not happen without hard work, and that satisfaction is the ultimate one. Talking about his organisation, he says, "Freshers start at about 120,000 rupees per annum, and an experience of a couple of years can fetch them anything between 180,000 and 300,000 rupees, depending on the field and competence of the person." Some reputed organisations offer as much as 220,000 rupees per annum to a fresh trainee engineer.

According to the data provided by TimesJobs.com, 50 per cent of the jobs are for entry-level engineers in the manufacturing domain, and they are paid anywhere between 150,000 and 250,000 rupees, depending on their skills and capabilities. Once they gain experience and elevate to junior-engineer level, their package can become anywhere between 350,000 and 700,000 rupees. Currently, about 26 per cent of the jobs are for fairly experienced, junior-level professionals. About 10 per cent of the jobs are for mid-level and senior-level experts each in the manufacturing domain.

Expert advice

Anand believes practise is the key. He says, "If you rely only on college/institutional knowledge and practicals, this field may not be a good option for you. People who are interested in making career in SMT and electronics manufacturing industry must show extra-curricular efforts (in these areas) to demonstrate their worth. Doing this ensures their faster progress and growth in industry, once they join."

Interested candidates need to be aligned to the requirements of the firms in India. Students should get more involved and try to identify the industry sector they wish to join, suggests Mitra. Giving an example of SMTA student chapter, he says,

MAJOR CONTRIBUTORS TO THIS ARTICLE

1. **Amarpreet Singh,** manager–manufacturing, Centum Electronics
2. **Amit Bhargava,** managing director, Asha Electronics
3. **Ankan Mitra,** vice-president, SMTA India Chapter
4. **Srinivasa Bharadwaj,** manager–engineering, Centum Electronics
5. **T. Anand,** principal consultant, Knewron

"SMTA's student chapters are currently located outside India. Within South East Asia, like Malaysia and Thailand, students take advantage of the local chapters to establish relation with the industry." Talking about India, he adds, "We are trying to establish student chapters in cooperation with some institutes, specifically in the southern part of the country. If students are interested, they can approach the SMTA contacts, and we can acquaint them with the industry requirements."

By 2020 jobs will hit mainstream

At present 75 per cent to 80 per cent of the electronics is imported and only about 20 per cent is manufactured in India, to fulfill its internal demand. So, obviously, there is a huge scope to manufacture within India. Mitra says, "This is the reason why all the top multinational electronics design or manufacturing companies have their offices in India already since the last decade or so. They are just waiting for the regulatory clearances and then they will expand in India the way they had initially planned to." With this being the situation, it definitely makes sense that people align themselves to this industry. Though the industry is picking up, it has not yet moved into that level of maturity. There will be tremendous job opportunities in the coming years, probably by 2020, besides those already existing today. ●

The author is a senior technical correspondent at EFY

Car-Reversing Audio-Visual Alarm

MD ARBAB ZAFAR DANISH

Here is a simple car-parking alarm circuit based on an AVR microcontroller and an ultrasonic module. The circuit will alert you while you are reversing your car for parking, if there is any obstacle, through an audio-visual alarm.

Circuit and working

Fig. 1 shows the circuit for the car-reversing alarm. It is built around ATmega328 AVR microcontroller (IC1), 5V regulator (IC2), ultrasonic module HC-SR04, piezo buzzer (PZ1) and a few other components.

ATmega328. The heart of the system is the 28-pin Atmel's ATmega328, which is an 8-bit AVR microcontroller with 32kB Flash and 1024 bytes of data RAM. It has two 8-bit timers/counters, one 16-bit timer/counter, six PWM channels, 23 programmable I/O lines, programmable serial USART master/slave SPI serial interface, a 6-channel 10-bit ADC and on-chip analogue comparator.

Pins 16 and 17 of IC1 are connected to Trigger and Echo pins of the ultrasonic module. Pin 18 of IC1 is connected to the piezo buzzer. The buzzer starts beeping when distance between the car's rear and an obstacle is 25cm or less.

Pins 4 through 6 and 11 through 15 of IC1 are connected to LED1 through LED8, respectively, to indicate distance between the car and the obstacle (refer Table I).

A 16MHz crystal oscillator is connected to pins 9 and 10 of IC1 to provide basic clock frequency. Power-on reset is provided by the combination of resistor R10 and capacitor C5. Switch S1 is used for manual reset of the microcontroller.

Ultrasonic transceiver HC-SR04. HC-SR04 ultrasonic transceiver module (Fig. 2) uses sonar to determine the distance of an object, like bats or dolphins do. It offers excellent non-contact range detection of 2cm to 400cm with high accuracy and stable readings in an

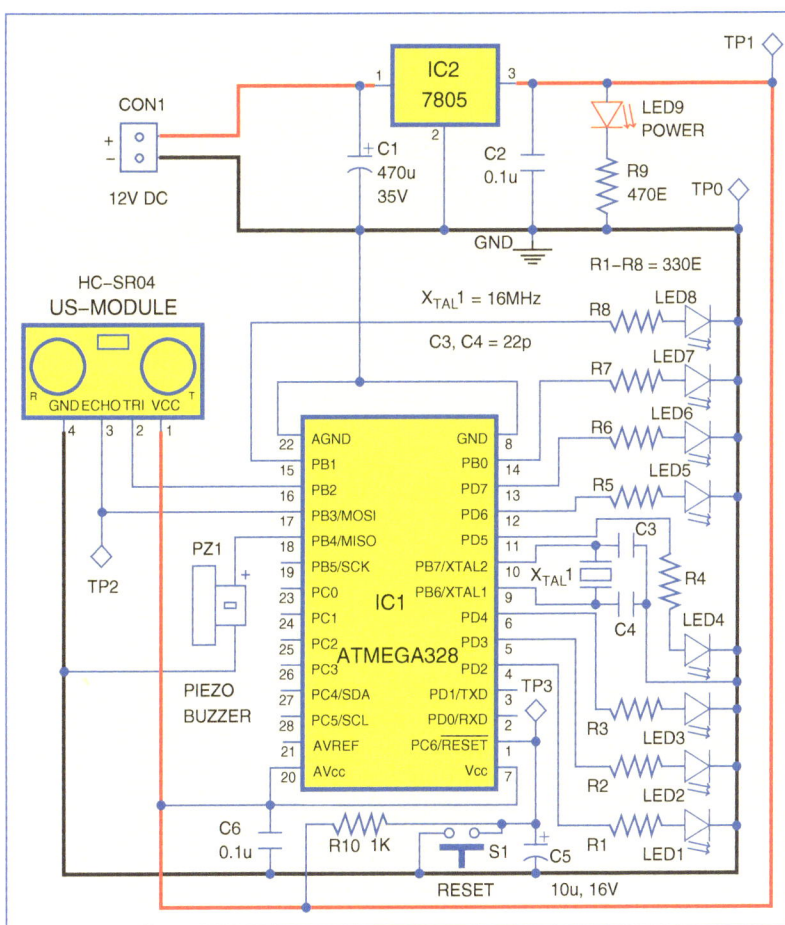

Fig. 1: Circuit diagram of the car-reversing audio-visual alarm

Fig. 2: Ultrasonic ranging module HC-SR04

TABLE I
LED Indications

Approx. distance from obstacle	LEDs glowing
200-175cm	LED1
175-150cm	LED1-LED2
150-125cm	LED1-LED3
125-100cm	LED1-LED4
100-75cm	LED1-LED5
75-50cm	LED1-LED6
50-25cm	LED1-LED7
25-0cm	LED1-LED8

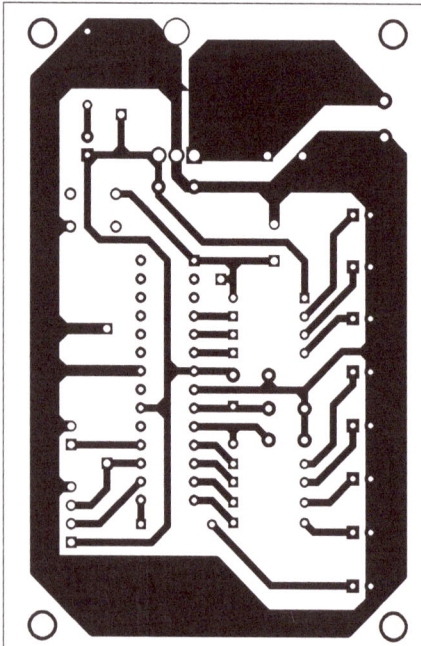

Fig. 3: An actual-size, single-side PCB for the car-reversing audio-visual alarm

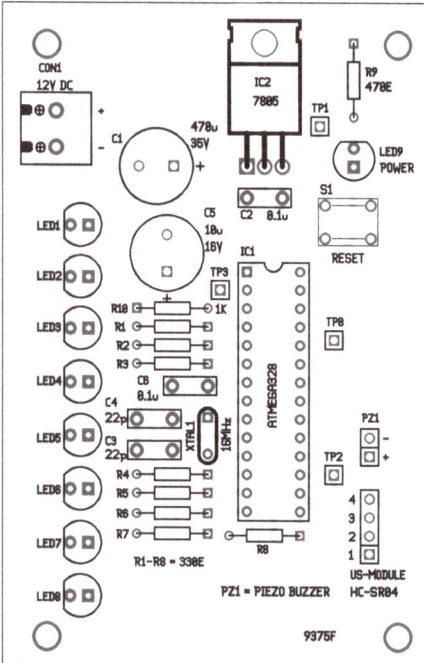

Fig. 4: Component layout for the PCB

easy-to-use package.

To determine the distance, pin 2 (TRI) of the module should receive a 'high' pulse for at least 10µs. This will initiate the module to transmit eight cycles of ultrasonic burst at 40kHz and wait for the reflected ultrasonic

signal. When the sensor detects the reflected ultrasonic signal, it sets pin 3 (Echo) to high. Duration of the reflected pulse depends on the distance from the obstacle. The distance can be calculated as:

Distance (in centimeters) = T/58, where T = Width of pulse at Echo pin in micro seconds

Power supply. 12V DC supply is drawn from the car's battery, or any other DC source, and converted to 5V using 7805 regulator, to power the circuit, including the microcontroller and sensor module HC-SR04. LED9 is used to indicate the presence of power in the circuit.

Software. The program is written in Arduino programming language and compiled and burnt using Arduino IDE. The software (parkingalarm.ino) is designed to detect an obstacle up to a maximum of 200cm distance.

Construction and testing

An actual-size, single-side PCB for the car-reversing audio-visual alarm circuit is shown in Fig. 3 and its component layout in Fig. 4.

Assemble the circuit on the recommended PCB to minimize assembly errors. Double-check for any overlooked error. Use proper IC base for the microcontroller. To test the circuit for proper functioning, verify correct 5V supply for the circuit at TP1 with respect to TP0.

For initial testing, just switch on the circuit and move an object in front of the sensor module. When distance between sensor and object is about 200cm, LED1 would glow. When the distance is about 25cm only, all the LEDs (LED1 through LED8) would glow and the buzzer will start giving a continuous beep sound.

You can install the sensor module at the rear bumper of your car, as shown in Fig. 5. The LEDs (LED1 through LED8) and buzzer can be placed near the dashboard. The circuit should get switched on as soon as you place your

Fig. 5: Suggested sensor installation at the rear bumper

TABLE II
Test Points

Test point	Details
TP0	0V, GND
TP1	5V
TP2	High pulse as per distance of obstacle
TP3	Low (when S1 is pressed)

PARTS LIST

Semiconductors:
IC1   - ATmega328 microcontroller
IC2   - 7805, 5V regulator
LED1-LED9   - 5mm LED
Resistors (all 1/4-watt, ±5% carbon):
R1-R8   - 330-ohm
R9   - 470-ohm
R10   - 1-kilo-ohm
Capacitors:
C1   - 470µF, 35V electrolytic
C2, C6   - 0.1µF ceramic disk
C3, C4   - 22pF ceramic disk
C5   - 10µF, 16V electrolytic
Miscellaneous:
PZ1   - Piezo-buzzer
$X_{TAL}1$ - 16MHz crystal oscillator
CON1   - 2-pin terminal connector
S1   - Tactile switch
US-Module   - HC-SR04 transceiver

EFY Note

The source code of this project is included in this month's EFY DVD and is also available for free download at source.efymag.com

car in reverse gear. As your car approaches an obstacle, more and more LEDs would start glowing. The buzzer will give a continuous beep sound when distance between your car and the obstacle becomes 25cm or less. ●

The author is a B.Tech (electronics and communication). His interests include working with microprocessors and microcontrollers

Power Factor Corrector

SUNIL KUMAR

This project discusses the need for power factor correction and provides a suitable DIY solution that could be used for small-scale industries and establishments.

When voltage and current are in phase with each other in an AC circuit, the electrical energy drawn from the source gets fully converted into another form of energy and the power factor (cosine of angle between voltage and current waveform) is said to be unity (or 100 per cent). This happens with purely resistive loads. With inductive loads, the voltage and current do not remain in phase and the power factor drops.

As the power factor drops, the system becomes less efficient. A drop from unity to 0.9 (90 per cent) in the power factor results in 15 per cent more current requirement for the same load. A power factor of 0.7 (70 per cent) requires approximately 43 per cent more current.

In industrial units and establishments, most of the loads are electrical motors and air-conditioning units. These loads are inductive in nature, where the current lags the applied voltage and the power factor is termed as lagging power factor. With capacitive loads, the current leads the voltage and the power factor is termed as leading power factor. The objective therefore should be to neutralise the lagging power factor of inductive loads by connecting capacitors across the load, which have leading power factor.

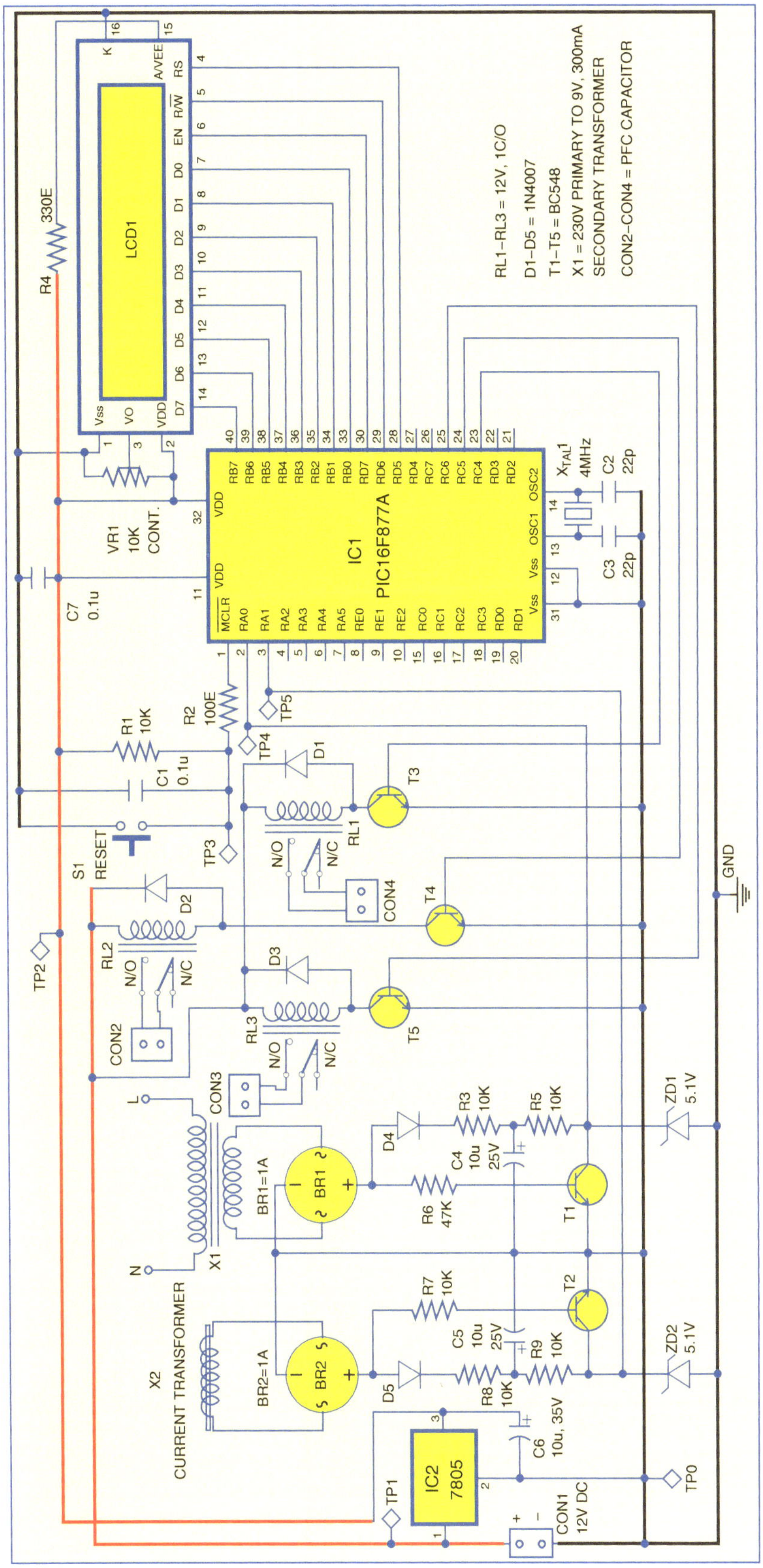

Fig. 1: The power factor correction circuit

TABLE I
Power Factor Multiplier

		Desired Power Factor				
		0.800	**0.850**	**0.900**	**0.950**	**1.000**
Original Power Factor	**0.50**	0.982	1.112	1.248	1.403	1.732
	0.55	0.768	0.899	1.034	1.190	1.518
	0.60	0.583	0.714	0.849	1.005	1.333
	0.65	0.419	0.549	0.685	0.840	1.169
	0.70	0.270	0.400	0.536	0.692	1.020
	0.75	0.132	0.262	0.398	0.553	0.882
	0.80	0.000	0.130	0.266	0.421	0.750
	0.85	—	0.000	0.135	0.291	0.620
	0.90	—	—	0.000	0.156	0.484
	0.95	—	—	—	0.000	0.329
	1.00	—	—	—	—	0.000

Configuration Bits				
✓ Configuration Bits set in code				
Address	Value	Field	Category	Setting
2007	3F39	OSC	Oscillator	XT
		WDT	Watchdog Timer	Off
		PUT	Power Up Timer	Off
		BODEN	Brown Out Detect	Off
		LVP	Low Voltage Program	Disabled
		CPD	Data EE Read Protect	Off
		WRT_ENABLE	Flash Program Write	Write Protection Off
		CP	Code Protect	Off

Fig. 2: Configuration bits

By improving the power factor you can save money on your electricity bill and also derive the following benefits:

1. Reduction of heating losses in transformers and distribution equipment

2. Longer equipment life

3. Increase in the capacity of your existing system and equipment.

While power factor correction is required for efficient use of electrical power, over correction of power factor is not recommended. In this project, over correction is not considered; we have considered power factors between 60 per cent and 90 per cent only due to inductive loads.

Circuit and working

Fig. 1 shows the circuit of a microcontroller-based power factor corrector. The circuit is built around PIC16F877A microcontroller (IC1), 230V AC primary to 9V, 300mA secondary transformer (X1), current transformer (X2), three relays (RL1-RL3), a 16x2 LCD display (LCD1) and a few other components.

Microcontroller PIC16F877A. Microcontroller PIC16F877A is the heart of the circuit. It is used to detect the phase difference between voltage and current in the AC mains supply line. It also connects the power factor correction (PFC) capacitor across the inductive load through a relay as explained below.

IC1 is a low-power, high-performance, CMOS 8-bit microcontroller. Its main features are 8kB Flash memory, 256-byte EEPROM, 368-byte RAM, 33 input/output (I/O) pins, 10-bit 8-channel analogue-to-digital converter (ADC), three timers, a watchdog timer with its own on-chip oscillator for reliable operation and synchronous I²C interface.

IC1's port pins RB0 through RB7 are connected to D0 though D7 of LCD1. Port pins RD5, RD6 and RD7 of IC1 are connected to control pins register select (RS), read/write (R/$\overline{\text{W}}$) and enable (EN) of LCD1, respectively. Port pins RC4 through RC6 are used to control relays RL1 through RL3, respectively.

In this project, three different PFC capacitors are used to connect across the inductive load through RL1, RL2 and RL3 relays. If power factor is between 60 per cent and 70 per cent, RL1 is energized. RL2 is energized if power factor is over 70 per cent but within 80 per cent and RL3 for power factor above 80 per cent but within 90 per cent.

When port pin RC4 of IC1 goes high, transistor T3 conducts and relay RL1 gets energized. This makes a PFC capacitor to connect across the inductive load to neutralise the lagging power factor. Similarly, port pin RC5 and RC6 can control relays RL2 and RL3 to neutralise the lagging power factor.

A 4MHz crystal oscillator is connected to pins 13 and 14 of IC1 to provide the basic clock frequency. Power-on reset is provided by the combination of resistor R1 and capacitor C1. Switch S1 is used for manual reset of the microcontroller. Port pins RA0 and RA1 of IC1 receive the zero-crossing detection pulses of voltage and current, respectively.

Zero-crossing detection. The zero-crossing is the instantaneous point in an AC waveform at which there is no voltage present. In this circuit, two zero-crossing detectors are employed to get the phase angle of voltage and current. First zero-crossing detector (comprising X1, BR1, T1 and ZD1) is used for detecting the point where the voltage crosses zero in either direction of the sinusoidal AC signals (positive and negative cycles).

The mains voltage of 230V AC is stepped down by transformer X1 to deliver a secondary output of 9V. The transformer output is rectified by full-wave bridge rectifier BR1 and applied to the base of transistor T1 through resistor R6. Capacitor C4 charges to its maximum value through diode D4 and provides supply to collector of transistor T1. Zener diode ZD1 regulates the voltage to 5.1V which is suitable for microcontroller input.

When rectified output transits through zero, T1 becomes off and its collector goes high. The detected voltage pulse is applied to port pin RA0 of IC1.

Similarly, zero-crossing detection for current is done through current transformer X2, bridge rectifier BR2, transistor T2 and zener diode ZD2. The current sample is rectified by bridge rectifier BR2. When rectified output goes through zero, a pulse is generated and applied to port pin RA1 of IC1.

The time difference between the voltage and the current pulses is calculated by the program embedded in the microcontroller. The power factor value is then calculated and displayed on the LCD. By knowing the value of power factor, we can calculate the value of PFC capacitor required.

PFC capacitor value. For neutralising the inductive reactance, we need to determine the value of PFC capacitor, including its kilo-volt-ampere-reactive (kVAR) rating, which needs to be

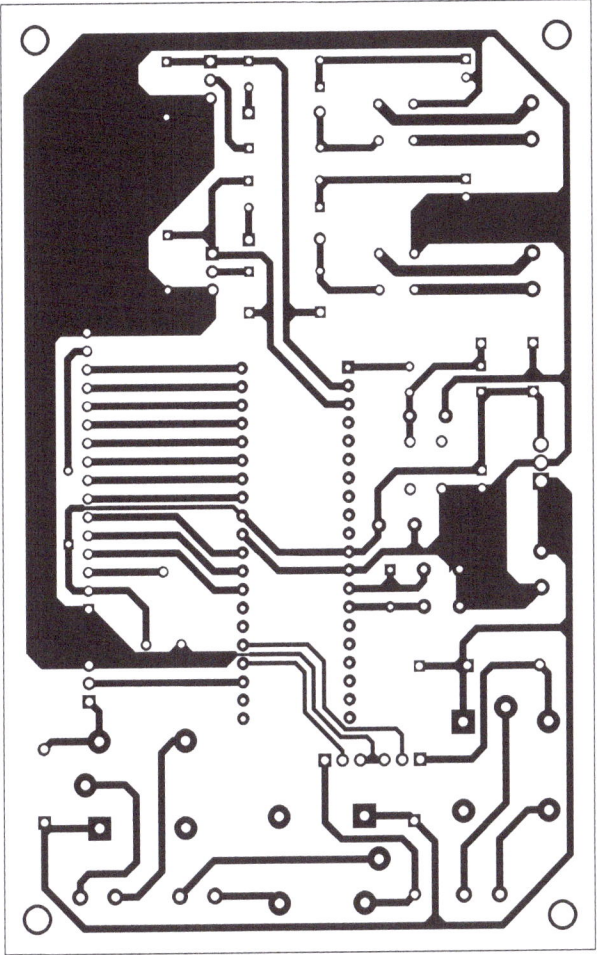

Fig. 3: An actual-size, single-side PCB for the PFC

Fig. 4: Component layout for the PCB

connected across the load. There are various methods to determine the corrective (leading) reactive power (kVAR) rating. Here we present a very simple method to know the kVAR in two steps:

1. Measure the RMS line voltage (line-to-neutral for single-phase load) and RMS line current. Power factor is already displaying on the LCD. For a single-phase system:

kW = volts × amps × PF/1000

2. Now determine the corrective (leading) reactive power (kVAR) needed for obtaining the desired PF by multiplying the kW value, as found above, with the selected value of power factor multiplier from Table I (intersection point of original PF in column 1 and the desired PF). For example, if the total plant load is 100kW at 60 per cent power factor. Capacitor

kVAR rating necessary to improve power factor to 90 per cent is found by multiplying kW (100) by the 0.849 multiplier in the table, which gives 84.9kVAR, or roughly 85kVAR. The standard PF capacitor rating nearest to 85kVAR should be used.

Software

The program is written in 'C' language and compiled using HI-TECH compiler along with MPLAB to generate hex code. The generated hex code is burnt into the microcontroller using suitable programmer with configuration bit setting as shown in Fig. 2. Program is well commented and easy to understand.

When the system is switched on, the main program initiates the LCD. Depending on the power factor across the load the 'PF is too low' message will be displayed in the first line and

power factor value, say, PF=56.70 per cent, in the second line of the LCD. Timer starts when a positive-going pulse is received at port pin RA0 and stops when port pin RA1 receives positive-going pulse. This timer count is converted to radians. Cosine value of the radians is the original power factor displayed on the LCD. One of the Port C pins (RC4, RC5 or RC6—depending on the power factor) goes high and energises the corresponding relay to connect the respective PFC capacitor across the load.

Construction and testing

An actual-size, single-side PCB for the PFC circuit is shown in Fig. 3 and its component layout in Fig. 4. Assemble the circuit on the recommended PCB to save time and minimize assembly errors. Carefully assemble the com-

PARTS LIST

Semiconductors:
IC1 — PIC16F877A microcontroller
IC2 — 7805, 5V regulator
BR1, BR2 — 1A bridge rectifier module
LCD1 — 16×2 LCD
D1-D5 — 1N4007 rectifier diode
T1-T5 — BC548 npn transistor
ZD1, ZD2 — 5.1V zener diode

Resistors (all 1/4-watt, ±5% carbon):
R1, R3, R5,
R7-R9 — 10-kilo-ohm
R2 — 100-ohm
R4 — 330-ohm
R6 — 47-kilo-ohm
VR1 — 10-kilo-ohm preset

Capacitors:
C1, C7 — 0.1µF ceramic
C2, C3 — 22pF ceramic
C4, C5 — 10µF, 25V electrolytic
C6 — 10µF, 35V electrolytic

Miscellaneous:
CON1-CON4 — 2-pin terminal connector
RL1-RL3 — 12V, 1 C/O relay
S1 — Tactile switch
X1 — 230V AC primary to 9V, 300mA secondary
X2 — Current transformer, 50/1A ratio
$X_{TAL}1$ — 4MHz crystal oscillator

EFY Note

The source code of this project is included in this month's EFY DVD and is also available for free download at source.efymag.com

ponents and double-check for any overlooked error. Use proper IC base for the microcontroller.

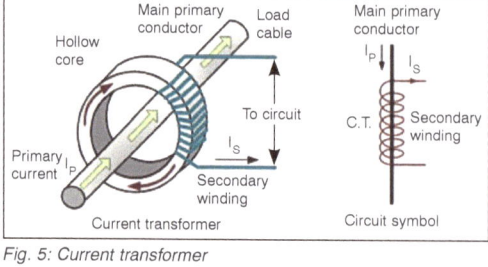

Fig. 5: Current transformer

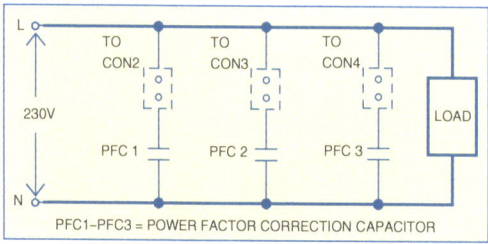

PFC1–PFC3 = POWER FACTOR CORRECTION CAPACITOR

Fig. 6: Block diagram of PFC capacitor connections

TABLE II
Test Points

Test point	Details
TP0	0V, GND
TP1	12V
TP2	5V
TP3	Low (when S1 is pressed)
TP4	High pulse (voltage zero-crossing detector)
TP5	High pulse (current zero-crossing detector)

Connect transformer X1 to the same AC mains supply where the load (motor) is connected. Connect the secondary terminals of current transformer X2 to the circuit. Power supply cable of the load should pass through the current transformer as shown in Fig. 5.

When you power on the circuit, the LCD will show the original or existing power factor value. You can find out the value of PFC capacitor as indicated above. The PFC capacitors and the load are to be connected externally. The PFC capacitor is connected across the load through the relay contact as shown in Fig. 6. Extend two-wire cable from the dotted line, as shown in the block diagram, to the respective connector on the PCB.

To test the circuit for proper functioning, verify the correct 12V and 5V supplies in the circuit at TP1 and TP2 with respect to TP0. Refer test points Table II for checking at various points in the circuit. ●

The author is a B.Tech in electronics and communication from AIT, Bengaluru

Alcohol Level Tester

EFY
S.C. DWIVEDI
TESTED

YOGESH SHUKLA

A lot of accidents happen every day due to drunk driving. This is a very useful circuit for testing whether a driver is drunk or not. The circuit is easy to use, inexpensive and indicates various levels of alcohol consumption through LEDs.

Circuit and working

Fig. 1 shows circuit of the alcohol level tester. Alcohol sensor MQ3 is used here to detect the alcohol fumes' concentration.

Pins 2 and 5 of MQ3 are connected to 5V power supply and the ground, respectively, as shown in the circuit. These pins are actually the heating-coil pins of the sensor. The input pins 1 and 3 of the sensor are also connected to 5V. Pins 4 and 6 are connected to input pin 5 of LM3914 (IC1).

The number of glowing LEDs indicates the concentration level of alcohol

PARTS LIST
Semiconductors:
IC1 — LM3914 bar graph
LED1-LED10 — Flat LED
Resistors (all 1/4-watt, ±5% carbon):
R1 — 500-ohm
R2 — 2.7-kilo-ohm
R3 — 3.9-kilo-ohm
VR1 — 20-kilo-ohm preset
Miscellaneous:
CON1 — 2-pin terminal connector
S1 — On/off switch
SENSOR1 — MQ3 alcohol sensor

Test Points

Test point	Details
TP0	0V
TP1	5V
TP2	MQ3 sensor output

detected by the sensor. Resistor R1 and preset VR1 are used to calibrate the output voltage from the sensor.

When a drunk breathes out near sensor1, the alcohol vapours come in contact with MQ3 sensor and its resistance changes. This raises signal level at pin 5 of IC1 a larger number of LEDs start glowing than the two that were glowing before.

Construction and testing

An actual-size, single-side PCB for the alcohol level tester is shown in Fig. 2 and its component layout in Fig. 3. After assembling the circuit on PCB, enclose it in a suitable case.

Switch on the circuit and verify the test points mentioned in the table. Before using the circuit, vary VR1 until LED1 and LED2 glow when the sensor is in normal environment. Now open the cork of an alcohol bottle and bring it near sensor1. You will notice that LED3, LED4 and LED5 also start glowing. As you bring the alcohol bottle's open mouth very close to sensor1, all the LEDs (LED1 through LED10) will start glowing.

We do not recommend that you actually consume some alcohol to test the circuit. The above test with a bottle of alcohol, or its sample in a small glass, should suffice for most purposes. Until you find a drunk, of course! ●

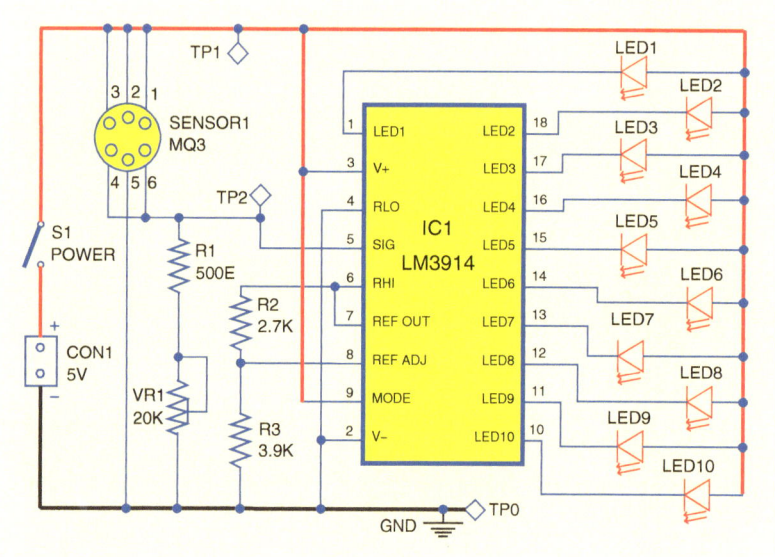

Fig. 1: The alcohol level tester's circuit diagram

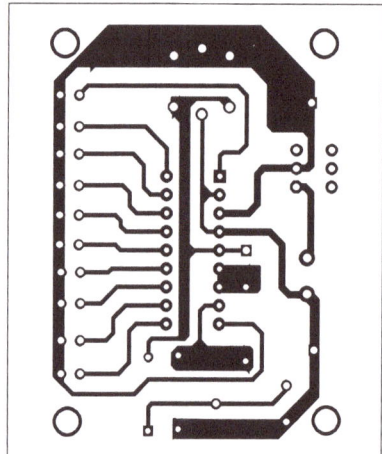

Fig. 2: An actual-size, single-side PCB layout for the circuit

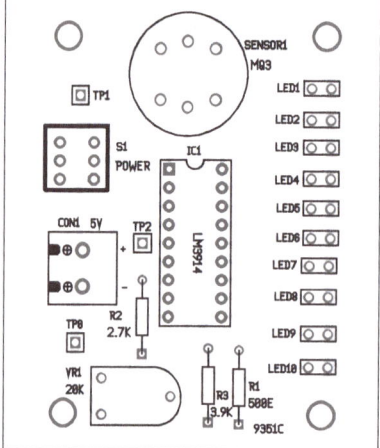

Fig. 3: Component layout for the PCB

The author is an electronics hobbyist

Brushless DC Motor Driver

ABHIJEET RAI

Use of brushless DC motors (BLDCs) is on the rise. But their control usually requires rotor-position information for selecting the appropriate commutation angle. Normally, a Hall Effect sensor is used to sense rotor position. But in cost-sensitive applications, a sensor-less commutation scheme is often desirable. The circuit described here uses a DRV10866 driver IC to drive a small BLDC fan, without using any position sensors. A BLDC fan's speed can be varied smoothly, without the usual steps associated with a normal AC fan.

Circuit and working

Fig. 1 shows the circuit of a sensor-less BLDC motor driver. The circuit is built around an NE555 (IC1), a DRV10866 (IC2) and a few other components.

DRV10866 driver IC from Texas Instruments is used to drive a small three-phase BLDC motor (M1). The circuit is of a three-phase, sensor-less motor driver with integrated power MOS-FETs having drive-current capability up to 680mA peak. DRV10866 is specifically designed for low noise and low component-count fan-motor drive applications. A 150° sensor-less back emf scheme is used to control the three-phase motor.

A 100k pull-up resistor (R2) is used at pin 1 of IC2. Pins 2, 4, 7 and 6 of IC2 are connected to common, phase A, phase B and phase C of the BLDC motor, respectively. Pin 10 of IC2 is connected to pin 7 of IC1 to get the pulse-width modulated (PWM) signal from IC1 to control the speed of the

PARTS LIST	
Semiconductors:	
IC1	- NE555 timer IC
IC2	- DRV10866, 3-phase BLDC motor driver
D1, D2	- BAT41 Schottky diode
Resistors (all 1/4-watt, ±5% carbon):	
R1	- 10-kilo-ohm
R2	- 100-kilo-ohm
R3	- 3.8-kilo-ohm
VR1	- 5-kilo-ohm potmeter
Capacitors:	
C1, C3	- 10nF ceramic disk (X7R recommended)
C2	- 0.1µF ceramic disk (X5R recommended)
C4	- 2.2µF ceramic disk (X7R recommended)
Miscellaneous:	
CON1	- 2-pin terminal block connector
M1	- 5V, 3-phase BLDC motor

Test Points

Test point	Details
TP0	0V
TP1	5V
TP2	PWM pulse

BLDC motor.

The output signal (PWM) is available at IC1's pin 7 (DIS) and not from the usual output pin 3 of the IC. The 25kHz (approx.) PWM signal's duty cycle can be adjusted from 5% to 95% using potentiometer VR1. The speed of the BLDC motor can be controlled by varying the duty cycle of the PWM signal. Turning VR1 counter-clockwise lowers the duty cycle which, in turn, lowers the speed of the motor, and vice versa.

Construction and testing

An actual-size, single-side PCB for the brushless DC motor driver is shown in Fig. 2 and its component layout in Fig. 3. Assemble the circuit on the recommended PCB to minimise assembly errors. IC2 should be fitted on solder side of the PCB.

After assembling the components, connect a 5V DC supply to CON1 connector. To test the circuit for proper functioning, verify correct 5V supply for the circuit at TP1 with respect to TP0. Turn VR1 clockwise or counter-clockwise to increase or decrease the speed of the motor. ●

The author is a B.Tech (electronics and communication) from GGSIPU, New Delhi

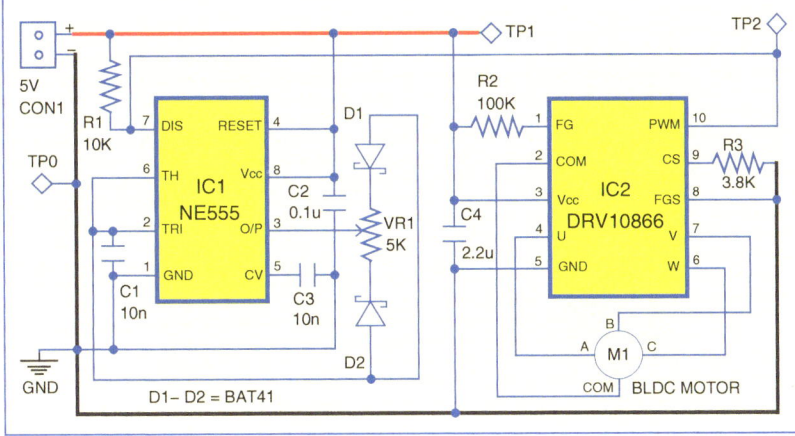

Fig. 1: Circuit of brushless DC motor driver

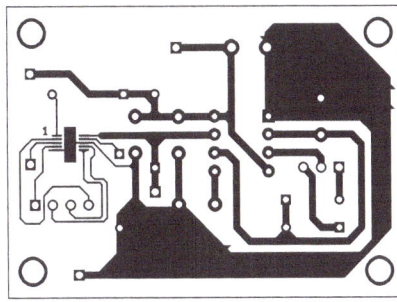

Fig. 2: An actual-size, single-side PCB for the brushless DC motor driver

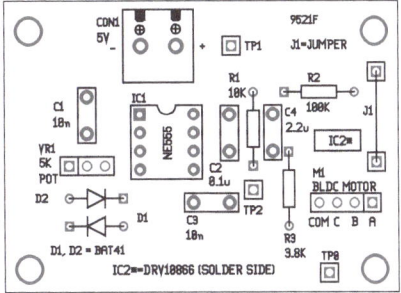

Fig. 3: Component layout for the PCB

Versatile Audio-Visual Alarm

PETRE TZV PETROV

This circuit uses an NE555 timer IC, some LEDs, a couple of piezo buzzers and a few other components to produce audio-visual effects as per your requirement. The timer NE555 and its equivalents are widely used for all sorts of audio and visual indications, such as door alarms. But the sound produced by these circuits may not be always pleasant to hear, or the light produced may not be visually appealing. With this circuit you can get different audio-visual effects.

Here we use LEDs for visual indication and buzzers for audible alarms as they require relatively low current to operate. By simply connecting some resistors and capacitors to NE555 we can obtain some interesting visual and audible effects as described here.

Circuit and working

Fig. 1 shows the circuit of the versatile audio-visual alarm which is built around timer NE555 (IC1), LEDs, buzzers and some resistors and capacitors. Resistors R1 and R2 and capacitor C1 determine the frequency of the LEDs' blinking. The frequency is selected usually within the range of 0.1Hz to 20Hz, depending on your requirement. Values of resistors R1 and R2 can be above 1-kilo-ohm. Capacitor C1's value can be between 1μF and 1000μF.

Timer NE555 drives two outputs, namely, Group1 and Group2. Group1 is built around resistors R4 and R6 along with LED1 through LED6. Group2 is built around resistors R7 and R8 along with LED7 through LED12.

Each of the groups can be configured to get different outputs. For example, in Group1 you can use only the LEDs (LED1 through LED3) connected to +12V, or only the LEDs (LED4 through LED6) connected to the ground, or both branches of these LEDs, or only piezo buzzer PZ1, or PZ1 with any combination of the LEDs, or you can omit the entire Group1.

The components in Group2 can form the same combinations as the components in Group1. The difference between the Group1 and Group2 is the use of resistor R5 and capacitor C2. These two components give light-decay effect to the LEDs and a pleasant low-pitch sound to piezo buzzer in Group2. Value of resistor R5 can be between 75-ohm and 1-kilo-ohm and that of capacitor C2 between 47μF and 1000μF.

At point 1 (TP2) in the circuit you can see a rectangular wave signal. At point 2 you can see a triangular or trapezoidal-like signal. The signals at points 1 and 2 should go low, almost to zero, and should go high, almost to 12V supply voltage.

Power supply used is 12V, but it

Test Points

Test point	Details
TP0	0V
TP1	12V
TP2	Rectangular wave
TP3	Triangular wave

PARTS LIST

Semiconductors:
IC1 — NE555 timer
LED1-LED13 — 5mm LED
D1 — 1N4148 signal diode

Resistors (all 1/4-watt, ±5% carbon):
R1, R3 — 10-kilo-ohm
R2 — 1-mega-ohm
R4, R6-R8 — 470-ohm
R5 — 100-ohm
R9 — 1-kilo-ohm

Capacitors:
C1 — 4.7μF, 25V electrolytic
C2 — 1000μF, 25V electrolytic
C3 — 0.33μF ceramic disk
C4 — 220μF, 25V electrolytic
C5 — 0.1μF ceramic disk

Miscellaneous:
CON1 — 2-pin terminal connector
CON2 — 2-pin connector
PZ1-PZ2 — Piezo buzzer with internal oscillators
S1 — On/off switch
— 8-pin IC base

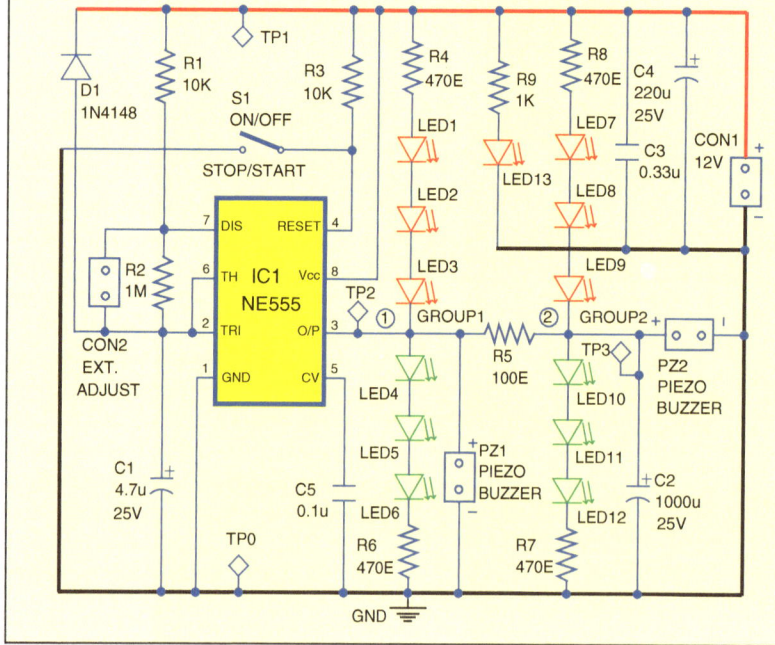

Fig. 1: The versatile audio-visual alarm circuit

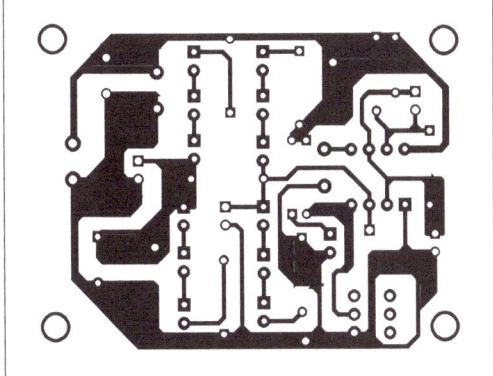

Fig. 2: Actual-size, single-side PCB for the circuit

Fig. 3: Component layout for the PCB

Construction and testing

An actual-size, single-side PCB for the versatile audio-visual alarm is shown in Fig. 2 and its component layout in Fig. 3. After assembling the circuit on PCB, enclose it in a suitable plastic box.

Connect piezo buzzers PZ1 and PZ2 at their provided places in the PCB. Also connect 2-pin terminal CON1 for power supply. Connect CON2 for external input (optional). Before using the alarm circuit, check at the test points given in the table. ●

can be in the range of 4.5V to 15V as well, depending on the number of LEDs used in each branch. Higher number of LEDs will require higher voltage. LED13 glows when power supply is connected in the circuit.

Resistors R4, R6, R7 and R8 are selected according to the number and type of the LEDs used. If the values of these resistors are too low, the output of the timer will be overloaded and the LEDs in the upper and the lower branches will get activated simultaneously.

Overloading may also damage the NE555 timer. It is suggested to keep the total output current drawn from NE555 below 100mA.

On/off switch S1 is used to start or stop the alarm. Connector CON2 is an optional input point for connecting a variable element, such as a preset, for adjusting or varying the frequency of the square signal for more audio-visual effects.

The author was a researcher and assistant professor in Technical University of Sofia (Bulgaria) and expert-lecturer in OFPPT (Casablanca), Kingdom of Morocco. Now he is working as an electronics engineer in the private sector in Bulgaria

Crystal-Controlled FM Transmitter

JOY MUKHERJI

Most simple FM transmitters use LC (inductor and capacitor) tuned circuits to generate the carrier waves. But they are inherently unstable due to drift of the resonant frequency. Even moving your body in front of the transmitter changes the frequency or destabilises it altogether. The crystal-controlled FM transmitter presented here is extremely stable. At the heart of the circuit is a phase-locked loop (PLL) clock multiplier chip ICS501.

Circuit and working

Fig. 1 shows the circuit diagram of crystal-controlled FM transmitter. It is built around 5V voltage regulator 7805 (IC1),

low-power audio amplifier LM386 (IC2), a PLL clock multiplier chip ICS501 (IC3), power transistor 2N3866A (T1) and a few other components.

The circuit works on 12V regulated power supply. IC1 provides 5V regulated voltage to drive IC3. The audio from the condenser microphone is amplified in audio amplifier stage built around IC2 and applied to the varactor diode D1 via resistor R3. The audio-frequency signal modulates the bias voltage on the varactor diode, varying the crystal's frequency and thus producing an FM signal at pin 5 of ICS501. This gives us about 40mW at the base of RF power transistor T1. The transistor T1 is biased in class C mode and provides a good RF gain. 50-ohm antenna-matching impedance is formed by inductor L2 and variable capacitors (also called trimmers) VC1 and VC2. Adjust L2, VC1 and VC2 for maximum output power.

The ICS501 uses 12MHz external reference crystal, which is multiplied by a factor of 8 (using S0 S1 as 11). The carrier level is around +27dBm (500mW) at 96MHz (8x12MHz). That is, when you speak into microphone MIC1, you will receive the transmitted

Test Points

Test point	Details
TP0	0V
TP1	12V when switch S1 is on
TP2	5V
TP3	Amplified audio signal
TP4	96MHz

PARTS LIST

Semiconductors:

IC1	- 7805, 5V voltage regulator
IC2	- LM386 low-power audio amplifier
IC3	- ICS501 PLL clock multiplier
D1	- MV209 epicap diode (varactor)
T1	- 2N3866A npn transistor

Resistors (all 1/4-watt, ±5% carbon, unless stated otherwise):

R1	- 10-kilo-ohm
R2	- 220-ohm
R3	- 56-kilo-ohm
R4	- 33-ohm
R5	- 10-ohm, 0.5W
R6	- 1-kilo-ohm
R7	- 220-ohm, 0.5W

Capacitors:

C1	- 470pF ceramic disk
C2, C5, C6, C8-C10	- 100nF ceramic disk
C3, C7	- 220µF, 25V electrolytic
C4	- 10µF, 25V electrolytic
VC1, VC2	- 50pF variable capacitor

Miscellaneous:

L1	- 12-turn air-core 8mm-dia 26SWG wire
L2	- 5-turn air-core 8mm-dia 26SWG wire
MIC1	- Electret condenser microphone
CON1	- 2-pin terminal connector
RFC1	- 10 to 20 turns on TV balun core
S1	- On/off switch
$X_{TAL}1$	- 12MHz crystal oscillator
ANT.	- Antenna around 78cm tall
	- Heatsink for T1

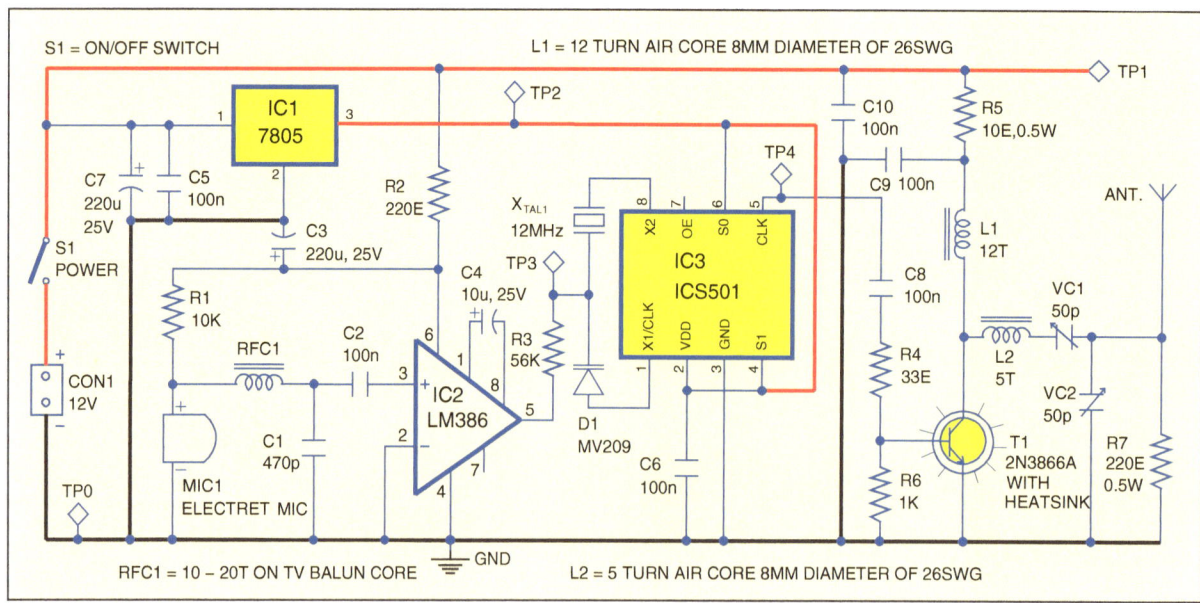

Fig. 1: Circuit diagram of crystal-controlled FM transmitter

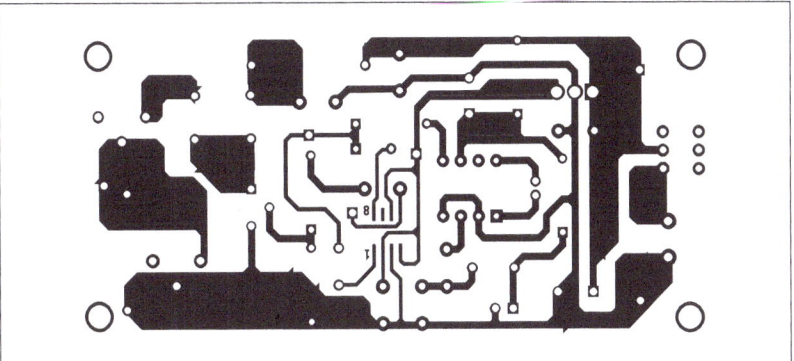

Fig. 2: An actual-size, single-side PCB for crystal-controlled single-frequency FM transmitter

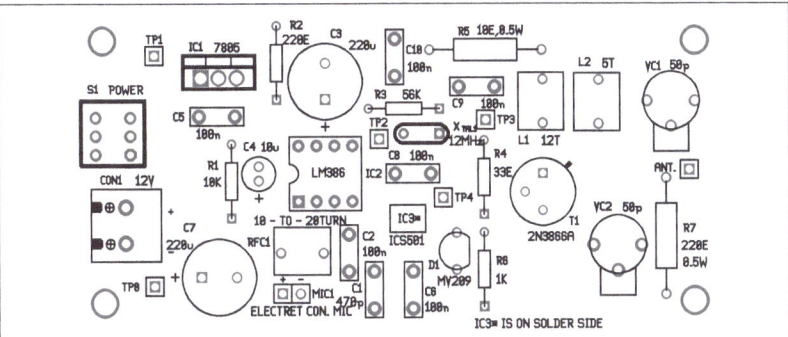

Fig. 3: Component layout for the PCB

signal in your FM receiver at 96MHz.

Construction and testing

An actual-size, single-side PCB for crystal-controlled single-frequency FM transmitter is shown in Fig. 2 and its component layout in Fig. 3. After assembling the circuit on a PCB, enclose it in a suitable plastic case.

Fix 2-pin connector CON1 for power supply. Use a suitable heatsink for transistor 2N3866A. Use a ¼-wavelength antenna, which should be around 78cm tall, for 96MHz.

You can construct the circuit even on a piece of copper-clad PCB. The copper cladding acts as the ground plane, giving an excellent performance. Pair the transmitter with a high-sensitivity receiver for a longer range of application.

Before using the circuit, verify the voltages and signal at the test points, as mentioned in the table. ●

The author is an electronics hobbyist and a small-business owner in Albany, New York, US. His interest includes designing radio-frequency circuits

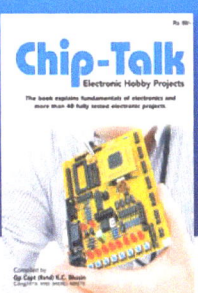

Simple IR Remote Control Extender

EFY
S.C. DWIVEDI
TESTED

RAJ K. GORKHALI

Sometimes you wish you could extend the range of the remote control of your TV set or DVD player, or take it to another room for use. This circuit can receive the IR signal from your remote control and retransmit it to the intended destination/equipment, even around a corner or to another room.

Circuit and working

Fig. 1 shows circuit diagram of the simple IR remote control extender. It is built around op-amp IC 741 (IC1), transistor CK100 (T1), photo-diode/receiving LED (IRRX1), transmitting LEDs (IRTX1 and IRTX2) and a few other components.

IRRX1 is basically a photo-diode, such as BPW41 or its equivalent BPW34. Since a photo-diode is designed to

PARTS LIST	
Semiconductors:	
IC1	- 741 op-amp
LED1	- 5mm LED
T1	- CK100 pnp transistor
IRTX1, IRTX2	- IR transmitting LED
IRRX1	- IR receiving LED (BPW34 or BPW41)
Resistors (all 1/4-watt, ±5% carbon):	
R1, R3, R4	- 100-kilo-ohm
R2	- 3.3-kilo-ohm
R5	- 1-mega-ohm
R6, R8	- 1-kilo-ohm
R7	- 56-ohm
VR1	- 470-ohm preset
Capacitors:	
C1	- 47nF ceramic disk
Miscellaneous:	
S1	- On/off switch
CON1	- 2-pin terminal connector

Test Points

Test point	Details
TP0	0V
TP1	9V
TP2	Train of pulses when IR signal is received

work in reverse bias mode, the cathode of IRRX1 is connected to positive supply rail (9V) through resistor R1. The non-inverting terminal of IC1 is held at half of the power supply by divider network comprising resistors R3 and R4.

The infrared signal from the remote control received at IRRX1 goes to pin 2 of IC1 through resistor R2 and capacitor C1. The output signal in the form of a train of pulses is available at IC1's output pin 6. This signal goes to transistor T1 through current-limiting resistor R8. Transistor T1 drives the IR transmitters IRTX1 and IRTX2. Thus the IR signal received at IRRX1 is amplified and retransmitted to the equipment through the IR transmitters.

The current flowing through IRTX1 and IRTX2 should not exceed 100mA. Hence a fixed current-limiting resistor (R7) is used in series with preset VR1. LED1, an ordinary LED, flashes when an IR signal is received from the remote for retransmission.

To control TV or air-conditioner from a longer distance or another room, connect the photodiode (IRRX1) with a pair of long wires as shown in Fig. 2.

Construction and testing

An actual-size, single-side PCB for the simple infrared control extender is shown in Fig. 3 and its component layout in Fig. 4. After assembling the circuit on PCB, enclose it in a suitable plastic case.

Fix a 2-pin connector (CON1) for power supply. Also fix a switch (S1) on the front panel for power on/off. Fix photo-diode IRRX1 in such a way that the remote's signal can easily fall on it. Keep the orientation of IR LEDs IRTX1 and IRTX2 such that the IR signals received at IRRX1 can be retransmitted like a repeater to a final destination through them.

To check the working of the circuit, point the remote towards IRRX1 and press any key of the remote. If LED1 flashes it means the IR signal is being received and getting retransmitted.

Now point IRTX1 and IRTX2 towards your equipment and again press any key of the remote. If you see LED1 flashing, you should be able to control your equipment through this circuit from a longer distance now.

Note. The directivity of the IR LEDs IRTX1 and IRTX2 and use of a reflector can increase the range further. ●

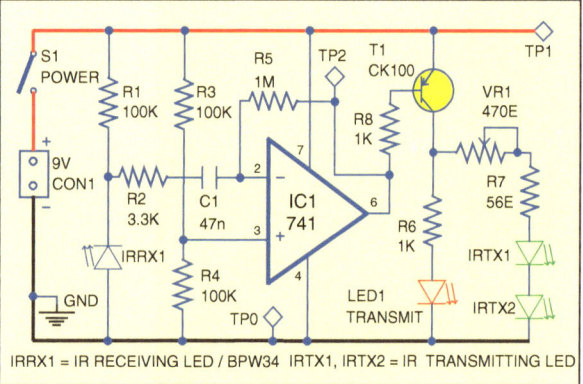

Fig. 1: Remote control extender's circuit diagram

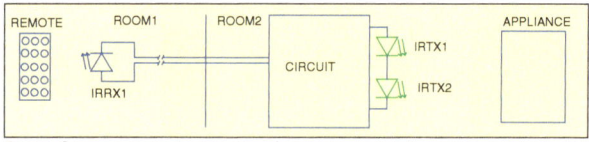

Fig. 2: Suggested wire extension of IR receiver between two rooms

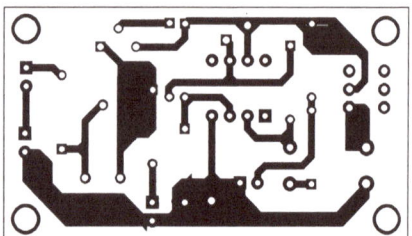

Fig. 3: Actual-size, single-side PCB layout for the circuit

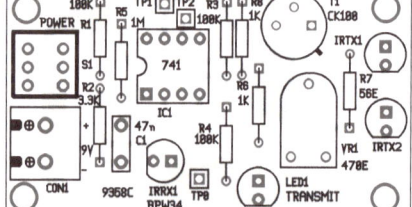

Fig. 4: Component layout for the PCB

The author is a regular contributor to EFY

Controlling Devices Through
WebIOPi Interface

SOMNATH BERA

In the evening, with the day's work over, your office is locked and you are about to leave for home. Suddenly, you notice that some of the lights in your office are still on, which you can see from the factory gate. You open the browser in your laptop, do authentication, give a command, and all the lights in your office are switched off. You return home peacefully.

Take another case. You have gone somewhere and realise that you will not be returning to your office for quite some time. So you go to a nearby computer, open the browser, check the temperature in your office, press some keys on the computer and the AC in your office gets switched off. You can do all this with the help of Raspberry Pi, using WebIOPi software.

WebIOPi

WebIOPi is browser-based software through which the above-mentioned tasks can be done easily yet economically. The latest WebIOPi version 0.0.6 is available at:

http://code.google.com/p/webiopi/downloads/detail?name=WebIOPi-0.6.0.tar.gz&can=2&q=

Download it with 'wget' command or directly from the browser. Extract the tarball file in a directory using the command (also refer Fig. 1):

```
$ tar zxvf WebIOPi-0.6.0.tar.gz
```

Now use 'cd' command to go to the extracted directory and run the following command (also refer Fig. 2):

```
$ sudo ./setup.sh
```

WebIOPi will be installed in your Raspberry Pi. To start WebIOPi, run the following command (also refer Fig. 3):

```
$ sudo /etc/init.d/webiopi start
```

This will not make WebIOPi start automatically at every boot. To start it automatically at the boot itself, give the command (also refer Fig. 4):

```
$ sudo update-rc.d webiopi defaults
```

Accessing Raspberry Pi's GPIOs

The WebIOPi is essentially an Apache webserver framework with facilities

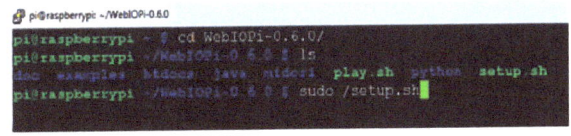

Fig. 1: Extracting the tarball file

Fig. 2: Installation of WebIOPi

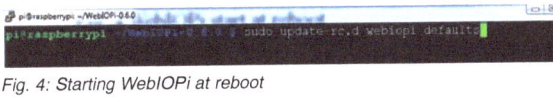

Fig. 3: Startig WebIOPi

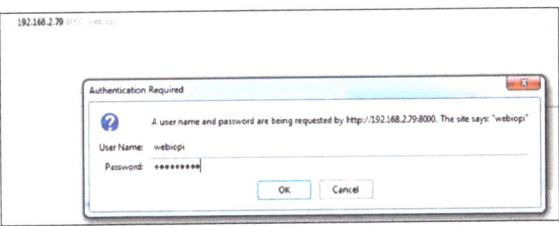

Fig. 4: Starting WebIOPi at reboot

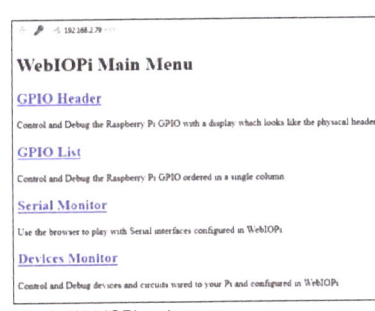

Fig. 5: Login to WebIOPi

WebIOPi Main Menu

GPIO Header

Control and Debug the Raspberry Pi GPIO with a display which looks like the physical header

GPIO List

Control and Debug the Raspberry Pi GPIO ordered in a single column

Serial Monitor

Use the browser to play with Serial interfaces configured in WebIOPi

Devices Monitor

Control and Debug devices and circuits wired to your Pi and configured in WebIOPi

Fig. 6: WebIOPi main menu

of Raspberry Pi GPIOs. If you already have Apache installed in your Raspberry Pi, just go ahead and WebIOPi will manage itself accordingly. If you do not have Apache, you can install it with the command:

```
$ sudo apt-get install apache2
```

Once WebIOPi is installed, go to the browser window of your computer and open the link:

```
http://your-raspi-
ip-address:8000/
webiopi
```

It will ask for a login and password, which you can change later. But for the time being type *webiopi* and *raspberry* as login and password, respectively, as shown in Fig. 5.

You will see the WebIOPi main menu, as shown in Fig. 6.

Click on the GPIO Header link and you will see the GPIO header pins along with the pin functions, as shown in Fig. 7. The 'IN' means the corresponding pin can be used as input. Double click on 'IN' box; the caption will change to 'OUT.' Now, the pin can be used as output pin.

Controlling devices

Make all the eight pins as OUT, then connect a few LEDs on the corresponding pins on the Raspberry Pi board. Click on each pin, say pin 13 (GPIO 27) to begin with, on the browser screen; it will change from black to orange and the corresponding LED will glow.

Connect all the GPIO pins to the inputs of a Darlington IC ULN2003 and you will be able to control relays connected to the outputs of the IC. The relays, in turn, will be able to switch ACs, fans, lights, curtains, etc on and off. As long as you can access this page with LAN or any other wireless network, you will be able to control the GPIO pins and the devices attached to them.

Similarly, click on 'OUT' and it will change to 'IN' button, indicating that the port has changed from an output port to an input port.

Interface temperature sensor

DS18B20 digital temperature sensor can sense temperature from -10 to 85 degree celsious. The sensor looks like a small transistor with its three legs, as shown in Fig. 8.

Fig. 8 also shows how to connect the GPIO of Raspberry Pi to temperature sensor DS18B20. Pin 1 (ground) of DS18B20 is connected to pin 6 (ground) of Raspberry Pi. Pin 2 (data) of DS18B20 is connected to pin 7 (GPIO4) of Raspberry Pi. GPIO4 is the only pin of Raspberry Pi which is used as general-purpose clock; any other GPIO pin will not work here. Pin 3 of DS18B20, which is for 3.3-volt supply, will go to pin 1 of Raspberry Pi, which has on-board 3.3-volt supply.

Ensure that a small resistor (4.7k to 10k) is connected between data pin and power pin, as shown in Fig. 8, to pull up data out signal for better correction. Switch on your Raspberry Pi and touch the sensor with a finger to see that it is not getting heated up excessively due to some wrong connection.

After login, load the DS18B20 module, and the temperature data can be viewed by using command (also refer Fig. 9):

```
$ sudo modprobe w1-gpio
$ sudo modprobe w1-therm
```

Write the command given below (also refer Fig. 10); you will find the ID number of the DS18B20 sensor as 28-000004ee2c8a:

```
$ ls /sys/bus/w1/devices/
```

Your temperature sensor's ID would be different but similar. Now open the webiopi config file using the command:

```
$ sudo nano /etc/webiopi/config
```

Go to the [DEVICE] section and add the following lines (also refer Fig. 11):

```
Attic-room-temp = DS18B20 slave:28-000004ee2c8a
tmp0 = DS18B20
```

The first line is for other than DS18B20 sensors, mostly the Chinese sensors. This way they will act as DS18B20 sensors. The second line (tmp0=DS18B20) is for DS18B20 sensors. It does not require any other argument. Both the commands will show the same temperature.

Next, reboot your Raspberry Pi, go to the WebIOPi main menu page and click on Devices Monitor link. You will see the temperature, as shown in Fig. 12.

Interface GPS receiver

You can even interface a GPS receiver and read the data from it on a web browser. Insert your GPS receiver into the USB port of Raspberry Pi and uncomment the following lines in the [DEVICES] section of the webiopi config file:

```
# USB serial adapters
usb0 = Serial device:ttyUSB0 baudrate:9600 //uncomment this line
```

Open the serial monitor on the WebIOPi menu and the GPS data starts pouring in. Just go through the /etc/webiopi/config file and you will come to know many other features of webiopi. ●

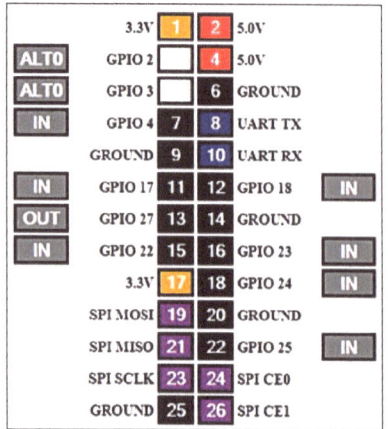

Fig. 7: GPIO header pins

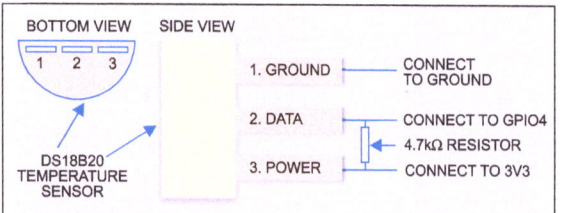

Fig. 8: Pin diagram of DS18B20

Fig. 9: Loading the DS18B20 module

Fig. 10: Finding the ID of temperature sensor

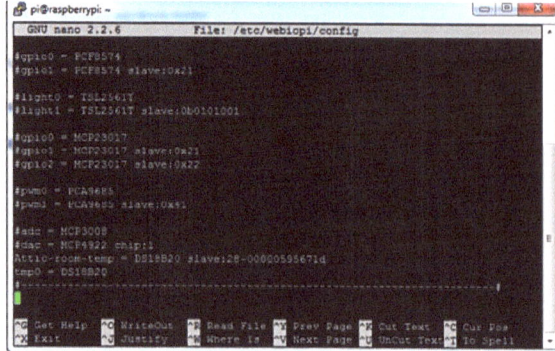

Fig. 11: WebIOPi config file

Devices Monitor

Attic-room-temp: Temperature

tmp0: Temperature: 28.94ºC

Fig. 12: Temperature display

The author is an avid user of open source software. Professionally, he is a thermal power expert and works as an additional general manager at NTPC Limited

Get Set Go with Raspberry Pi Camera

POOJA JUYAL

This project describes how to connect a web camera to the Raspberry Pi and make use of it. Raspberry Pi board has an onboard connector which can be used for connecting the camera directly. Connect the camera very carefully, as shown in Fig. 1, and take good grounding measures before touching it.

Installation of Raspberry Pi camera

First of all, open the Raspberry Pi's configuration file, Raspi config, with the command:

```
$ sudo raspi-config
```

Then enable the camera, as shown in Fig. 2.

If your firmware is old, update and upgrade it using the command:

```
$ sudo apt-get update
$ sudo apt-get upgrade
```

Now you should see the camera option. Enable it, as indicated in Fig. 2. After connecting the camera to the Raspberry Pi, see how best you can make use of it.

The command 'raspistill' is used to take still pictures. It comes with various parameters which you can choose for the command. Type raspistill in the terminal and you will see all the options, as shown in Table I.

For example, if you want to capture an image, type the command given below (also refer Fig. 3):

```
$ raspistill -v -o image.jpg
```

Here '-v' indicates that the output is verbose. You will see a preview of the snapshot for five seconds and the snapshot will be saved as image.jpg in the home directory. You can do several things with the image, such as defining its dimensions and quality. Check all the options and try them.

For video, you can use 'raspivid' command. For checking various options, type raspivid in the terminal. You will find all the options, as shown in Table II.

To capture a 10-second video with the Raspberry Pi camera, type the command (also refer Fig. 4):

```
$raspivid -o video.h264 -t 10000
```

Here 'video' is the name of your video, '.h264' is the file extension and '10000' is the time in milliseconds.

Check all the functions; you can do several things, such as recording the video and defining its frame size. You can even stream the video captured by the camera. The simplest method that I found for this was with mplayer and netcat.

Download both mplayer and netcat on a Windows computer from the link:

http://dl.dropboxusercontent.com/u/106074492/mplayer%20and%20netcat.zip

Extract files and you will see two exe files (mplayer.exe and nc.exe) that can be executed through command.

Install netcat and mplayer on Raspberry

Fig. 1: Raspberry Pi camera

```
info                 Information about this tool
expand_rootfs        Expand root partition to fill SD card
overscan             Change overscan
configure_keyboard   Set keyboard layout
change_pass          Change password for 'pi' user
change_locale        Set locale
change_timezone      Set timezone
change_hostname      Set hostname
memory_split         Change memory split
overclock            Configure overclocking
ssh                  Enable or disable ssh server
boot_behaviour       Start desktop on boot?
camera               Enable/Disable camera addon support

        <Select>              <Finish>
```

Fig. 2: Enabling the camera

```
pi@raspberrypi ~
pi@raspberrypi ~ $ raspistill -v -o image.jpg
```

Fig. 3: Capturing an image

```
pi@raspberrypi /
pi@raspberrypi / $ raspivid -o video.h264 -t 10000
```

Fig. 4: Recording the video

TABLE I
Raspistill Options

Commands	Functions
Image Parameter Commands	
-w, --width	Set image width <size>
-h, --height	Set image height <size>
-q, --quality	Set jpeg quality <0 to 100>
-r, --raw	Add raw bayer data to jpeg metadata
-o, --output	Output filename <filename> (to write to stdout, use '-o -'). If not specified, no file is saved
-v, --verbose	Output verbose information during run
-t, --timeout	Time (in ms) before it takes picture and shuts down (if not specified, set to 5s)
-sh, --sharpness	Set image sharpness (-100 to 100)
-co, --contrast	Set image contrast (-100 to 100)
-br, --brightness	Set image brightness (0 to 100)
-sa, --saturation	Set image saturation (-100 to 100)
-rot, --rotation	Set image rotation (0-359)
-hf, --hflip	Set horizontal flip
-vf, --vflip	Set vertical flip
Preview Parameter Commands	
-p, --preview	Preview window settings <'x,y,w,h'>
-f, --fullscreen	Full-screen preview mode
-op, --opacity	Preview window opacity (0-255)
-n, --nopreview	Do not display a preview window

TABLE II
Raspivid Options

Commands	Functions
Image Parameter Commands	
-w, --width	Set image width <size>, Default 1920
-h, --height	Set image height <size> Default 1080
-b, --bitrate	Set bitrate. Use bits per second (for example, 10MBits/s would be -b 10000000)
-o, --output	Output filename <filename> (to write to stdout, use '-o -'). If not specified, no file is saved
-v, --verbose	Output verbose information during run
-t, --timeout	Time (in ms) before it takes picture and shuts down (if not specified, set to 5s)
-sh, --sharpness	Set image sharpness (-100 to 100)
-co, --contrast	Set image contrast (-100 to 100)
-br, --brightness	Set image brightness (0 to 100)
-sa, --saturation	Set image saturation (-100 to 100)
-rot, --rotation	Set image rotation (0-359)
-hf, --hflip	Set horizontal flip
-vf, --vflip	Set vertical flip
Preview Parameter Commands	
-p, --preview	Preview window settings <'x,y,w,h'>
-f, --fullscreen	Full-screen preview mode
-op, --opacity	Preview window opacity (0-255)
-n, --nopreview	Do not display a preview window

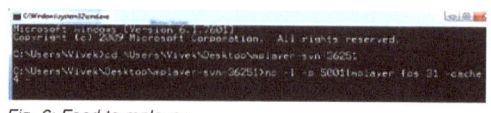

Fig. 5: Installing mplayer and netcat

Fig. 6: Feed to mplayer

Fig. 7: Sending the video to port 5001

Pi using the command (also refer Fig. 5):

```
$ sudo apt-get install mplayer netcat
```

You need to set up the Windows computer to receive transmission from the Raspberry Pi. So open command prompt and go to the directory where you have extracted mplayer and netcat. Type:

```
nc -l -p 5001|mplayer fps 31 -cache 1024 -
```

This will direct the feed from port 5001 to mplayer (refer Fig. 6).

Find the IP address of your Windows computer with ipconfig in command prompt.

Then go back to Raspberry Pi and type the command (also refer Fig. 7):

```
Raspivid -t 999999 -o - |nc [yourIP] 5001
```

This will take the video and send it to port 5001

When you go back to the computer, you will see the command prompt. Hit Enter key and the video starts streaming. You will see mplayer icon at the bottom-right corner of the window. Click it and you can see live footage. ●

The author is working as an assistant manager at Samtel ⬚vionics ⬚d

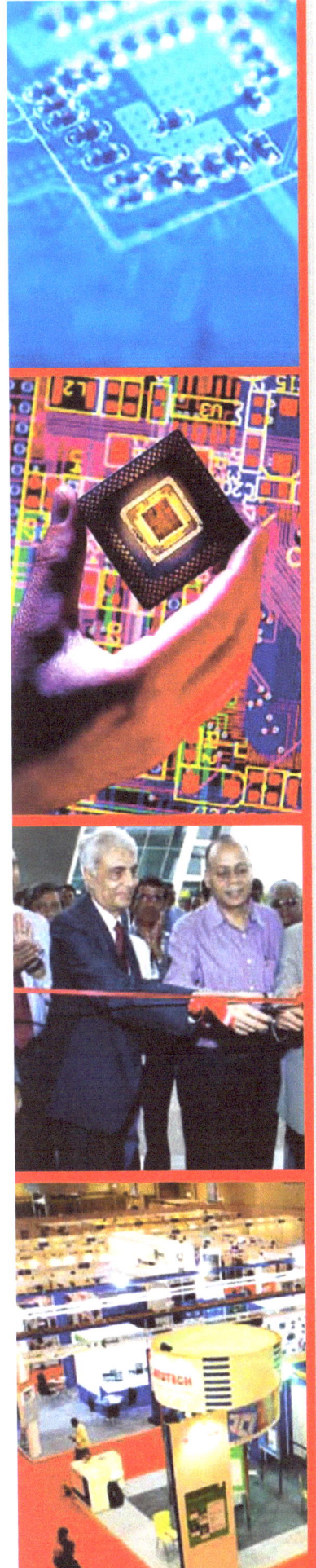

INDIAN PRINTED CIRCUIT ASSOCIATION

Pune's Biggest Buyers of Electronics are from the Automobile and Education Sectors

...and here is your chance to meet them

IPCA-EFYexpo 2014
August 6-8, Pune

Auto Cluster Exhibition Centre, Pune

Exhibit Today & Network With The Top Shots Of Pune's Manufacturing Industry

Call: Arun 08800094213 • **Email:** efyexpo@efyindia.com

Co-Partner By: • ELCINA • MAIT • SMTA • CEAMA • TEMA • CLIK • IESA

Other Co-Partners

KPCA	JPCA	HKPCA	CPCA	TPCA	EIPC	IPC
KPCA	JPCA	HKPCA	CPCA	TPCA	EIPC	IPC
Korea Printed Circuit Association, Seoul, Korea	Japan Printed Circuit Association, Tokyo, Japan	Hong Kong Printed Circuit Association, Hong Kong	China Printed Circuit Assocaltion, Shanghai, China	Taiwan Printed Circuit Association	European institute of printed Circuits	USA

ASSOCIATION CONNECTING ELECTRONICS INDUSTRIES

HCT: The HDL Complexity Tool

SHAKTHI KANNAN

HCT stands for HDL Complexity Tool, where HDL stands for Hardware Description Language. HCT provides scores that represent the complexity of modules present in integrated circuit (IC) designs. It is written in Perl and released under the GPLv3 and LGPLv3 license. It employs McCabe Cyclomatic Complexity that uses the control flow graph of the program source code to determine the complexity.

There are various factors for measuring the complexity of HDL models, such as size, nesting, modularity and timing. The measured metrics can help designers in refactoring their code, and also help managers to plan project schedules, and allocate resources, accordingly. You can run the tool from the Linux terminal for Verilog, VHDL and CDL (computer design language) files.

HCT can be installed on Fedora using the command:

```
$ sudo yum install hct
```

After installation, consider the example project of uart2spi written in Verilog, which is included in this month's EFY DVD. It implements a simple core for a UART interface and an internal SPI bus. The uart2spi folder contains rtl/spi under the file directory in your PC: /home/guest/uart2spi/trunk/rtl/spi

Run the HCT tool on the rtl/spi Verilog source as follows:

```
$ hct rtl/spi
```

We get the output:

```
Directory: /home/guest/uart2spi/trunk/rtl/spi
verilog, 4 file(s)
```

FILENAME	MODULE	IO	NET	MCCABE	TIME
spi_ctl.v		20	1	1	0.1724
	spi_ctl	20	1	1	
spi_core.v		0	0	1	0.0076
	spi_core	0	0	1	
spi_cfg.v		0	0	1	0.0076
	spi_cfg	0	0	1	
spi_if.v		15	3	1	0.0994
	spi_if	15	3	1	

The output includes various attributes that are described below:

1. FILENAME is the file that is being parsed. The parser uses the file name extension to recognize the programming language.

2. MODULE refers to the specific module present in the file. A file can contain many modules.

3. IO refers to the input/output registers used in the module.

4. NET includes the network entities declared in the given module. For Verilog, it can be 'wire,' 'tri,' 'supply0,' etc.

5. MCCABE provides the McCabe Cyclomatic Complexity of the module or file.

6. TIME refers to the time taken to process the file.

A specific metric can be excluded from the output using the "--output-exclude=LIST" option. For example, type the following command on a Linux terminal:

```
$ hct --output-exclude=TIME rtl/spi
```

The output will be:

```
Directory: /home/guest/uart2spi/trunk/rtl/spi
verilog, 4 file(s)
```

FILENAME	MODULE	IO	NET	MCCABE
spi_ctl.v		20	1	1
	spi_ctl	20	1	1
spi_core.v		0	0	1
	spi_core	0	0	1
spi_cfg.v		0	0	1
	spi_cfg	0	0	1
spi_if.v		15	3	1
	spi_if	15	3	1

If you want only the scores to be listed, you can remove the MODULE listing with the "--output-no-modules" option:

```
$ hct --output-no-modules rtl/spi
Directory: /home/guest/uart2spi/trunk/rtl/spi
verilog, 4 file(s)
```

FILENAME	IO	NET	MCCABE	TIME
spi_ctl.v	20	1	1	0.16803
spi_core.v	0	0	1	0.007434
spi_cfg.v	0	0	1	0.00755
spi_if.v	15	3	1	0.097721

The tool can be run on individual files, or recursively on subdirectories with the "-R" option. The output of the entire uart2spi project sources is given below:

```
$ hct -R rtl
Directory: /home/guest/uart2spi/trunk/rtl/uart_core
verilog, 4 file(s)

+---------------+-----------+------+-----+--------+----------+
| FILENAME      | MODULE    | IO   | NET | MCCABE |  TIME    |
+---------------+-----------+------+-----+--------+----------+
| uart_rxfsm.v  |           | 10   | 0   | 1      | 0.1379   |
|               | uart_rxfsm| 10   | 0   | 1      |          |
+---------------+-----------+------+-----+--------+----------+
| clk_ctl.v     |           | 0    | 0   | 1      | 0.0146   |
|               | clk_ctl   | 0    | 0   | 1      |          |
+---------------+-----------+------+-----+--------+----------+
| uart_core.v   |           | 18   | 1   | 1      | 0.1291   |
|               | uart_core | 18   | 1   | 1      |          |
+---------------+-----------+------+-----+--------+----------+
| uart_txfsm.v  |           | 9    | 0   | 1      | 0.1129   |
|               | uart_txfsm| 9    | 0   | 1      |          |
+---------------+-----------+------+-----+--------+----------+


Directory: /home/guest/uart2spi/trunk/rtl/top
verilog, 1 file(s)

+---------------+-----------+------+-----+--------+----------+
| FILENAME      | MODULE    | IO   | NET | MCCABE |  TIME    |
+---------------+-----------+------+-----+--------+----------+
| top.v         |           | 16   | 0   | 1      | 0.0827   |
|               | top       | 16   | 0   | 1      |          |
+---------------+-----------+------+-----+--------+----------+


Directory: /home/guest/uart2spi/trunk/rtl/spi
verilog, 4 file(s)

+---------------+-----------+------+-----+--------+----------+
| FILENAME      | MODULE    | IO   | NET | MCCABE |  TIME    |
+---------------+-----------+------+-----+--------+----------+
| spi_ctl.v     |           | 20   | 1   | 1      | 0.1645   |
|               | spi_ctl   | 20   | 1   | 1      |          |
+---------------+-----------+------+-----+--------+----------+
| spi_core.v    |           | 0    | 0   | 1      | 0.0074   |
|               | spi_core  | 0    | 0   | 1      |          |
+---------------+-----------+------+-----+--------+----------+
| spi_cfg.v     |           | 0    | 0   | 1      | 0.0073   |
|               | spi_cfg   | 0    | 0   | 1      |          |
+---------------+-----------+------+-----+--------+----------+
| spi_if.v      |           | 15   | 3   | 1      | 0.0983   |
|               | spi_if    | 15   | 3   | 1      |          |
+---------------+-----------+------+-----+--------+----------+


Directory: /home/guest/uart2spi/trunk/rtl/lib
verilog, 1 file(s)

+---------------+---------------+----+-----+--------+----------+
| FILENAME      | MODULE        | IO | NET | MCCABE |  TIME    |
+---------------+---------------+----+-----+--------+----------+
| registers.v   |               | 5  | 0   | 1      | 0.0382   |
|               | bit_register  | 5  | 0   | 1      |          |
+---------------+---------------+----+-----+--------+----------+
```

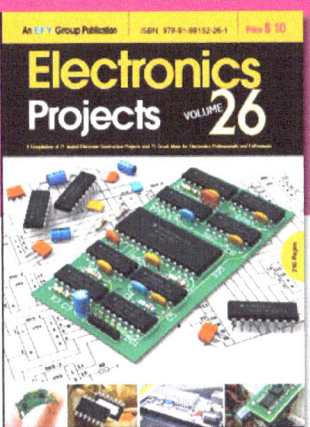

```
Directory: /home/guest/uart2spi/trunk/rtl/msg_hand
verilog, 1 file(s)

+--------------------+------------+----+-----+--------+--------+
| FILENAME           | MODULE     | IO | NET | MCCABE | TIME   |
+--------------------+------------+----+-----+--------+--------+
| uart_msg_handler.v |            | 0  | 0   | 1      | 0.0192 |
|                    | uart_m~ndler| 0 | 0   | 1      |        |
+------------------------------------------------------------+
```

The default behaviour is to dump the output to the terminal. It can be redirected to a file with the "- -output-file=FILE" option. You can also specify an output file format, such as "csv" with the "- -output-format=FORMAT" option:

```
$ hct --output-file=/home/guest/project-metrics.csv
--output-format=csv rtl/spi
$ cat /home/guest/project-metrics.csv
```

```
Directory: /home/guest/uart2spi/trunk/rtl/spi
verilog, 4 file(s)
FILENAME  , MODULE   , IO , NET , MCCABE , SLOC , COMMENT_LINES , TIME
spi_ctl.v ,          , 20 ,1   ,1      , 110 , 48            , 0.1644
          , spi_ctl  , 20 ,1   ,1      , 68  , 6             ,
spi_core.v,          , 0  ,0   ,1      , 46  , 43            , 0.0073
          , spi_core , 0  ,0   ,1      , 4   , 1             ,
spi_cfg.v ,          , 0  ,0   ,1      , 46  , 43            , 0.0075
          , spi_cfg  , 0  ,0   ,1      , 4   , 1             ,
spi_if.v  ,          , 15 ,3   ,1      , 80  , 44            , 0.0948
          , spi_if   , 15 ,3   ,1      , 38  , 2             ,
```

There are various yyparse options that are helpful to understand the lexical parsing of the source code. They can be invoked using following command:

```
$ hct --yydebug=NN sources
```

The NN options and their meaning is listed below:

0x01 Lexical tokens
0x02 Information on States
0x04 Shift, reduce, accept driver actions
0x08 Dump of the parse stack
0x16 Tracing for error recovery
0x31 Complete output for debugging

HCT can also be used with VHDL and Cyclicity CDL (cycle description language) programs. For VHDL, the filenames must end with a .vhdl extension. You can rename .vhd files recursively in a directory (in bash, for example) using the following script:

```
for file in `find $1 -name "*.vhd"`
do
mv $file ${file/.vhd/.vhdl}
done
```

The "$1" refers to the project source directory that is passed as an argument to the script.

Let us take example of sha256core written in VHDL, which is also included in this month's EFY DVD. The execution of HCT on the sha256core project is as follows:

```
$ hct rtl
Directory: /home/guest/sha256core/trunk/rtl
vhdl, 6 file(s)
```

```
+---------------+------------+----+-----+--------+----------+
| FILENAME      | MODULE     | IO | NET | MCCABE | TIME     |
+---------------+------------+----+-----+--------+----------+
| sha_256.vhdl  |            | 29 | 0   | 1      | 0.9847   |
|               | sha_256    | 29 | 0   | 1      |          |
+------------------------------------------------------------+
| sha_fun.vhdl  |            | 1  | 1   | 1      | 0.3422   |
|               |            | 1  | 1   | 1      |          |
+------------------------------------------------------------+
| msg_comp.vhdl |            | 20 | 0   | 1      | 0.4169   |
|               | msg_comp   | 20 | 0   | 1      |          |
+------------------------------------------------------------+
| dual_mem.vhdl |            | 7  | 0   | 3      | 0.0832   |
|               | dual_mem   | 7  | 0   | 3      |          |
+------------------------------------------------------------+
| ff_bank.vhdl  |            | 3  | 0   | 2      | 0.0260   |
|               | ff_bank    | 3  | 0   | 2      |          |
+------------------------------------------------------------+
| sh_reg.vhdl   |            | 19 | 0   | 1      | 0.6189   |
|               | sh_reg     | 19 | 0   | 1      |          |
+------------------------------------------------------------+
```

The "-T" option enables the use of threads to speed up computation. The LZRW1 (Lempel–Ziv Ross Williams) compressor core project implements a lossless data compression algorithm. The output of HCT on this project, without threading and with threads enabled, is shown below:

```
$ time hct HDL
Directory: /home/guest/lzrw1-compressor-core/trunk/hw/HDL
vhdl, 8 file(s)
...
real    0m3.725s
user    0m3.612s
sys     0m0.013s


$ time hct HDL -T

Directory: /home/guest/lzrw1-compressor-core/trunk/hw/HDL
vhdl, 8 file(s)
...
real    0m2.301s
user    0m7.029s
sys     0m0.051s
```

The supported input options for HCT can be viewed with the "-h" command.

The invocation of HCT can be automated, rechecked for each code check-in that happens to a project repository. The complexity measure is thus recorded periodically. The project team will then be able to monitor, analyse the complexity of each module and decide on any code refactoring strategies. ●

EFY Note

The source code of this project is included in this month's EFY DVD and is also available for free download at source.efymag.com

The author is a software enthusiast

Part 5 of 5
Designing with FPGAs:
FPGA-Embedded Processors

VARSHA AGRAWAL

FPGA devices are suitable for implementing parallel algorithms. However, sequential algorithms, especially those that do not demand large processing power, are easier to implement as a program in a microprocessor. Many applications require both microprocessor and an FPGA array. One possible way to implement this is to have separate CPU and FPGA chips. But the better approach is to combine them into a single chip. This reduces the power consumption, leads to simple board layout and fewer problems with signal integration and EMI.

This article gives information on FPGA-embedded processors with particular emphasis on the Embedded Development Kit (EDK) from Xilinx. The article also presents an illustrative example of a hands-on experience with the EDK design.

FPGA-embedded processors

There are two types of CPU cores for FPGAs, namely, hard and soft. Hard CPU core is a dedicated part of the integrated circuit, whereas a soft CPU is implemented utilising general-purpose FPGA logic cells.

Examples of FPGA chips with hard CPU cores include Virtex-4FX and Virtex-5 FXT series with PowerPC cores, and Atmel's FPSLIC with an AVR core. Some of the soft CPU cores include MicroBlaze and PicoBlaze that are used for the FPGAs manufactured by Xilinx, and Nios II processors that are limited to Altera devices.

MicroBlaze has a proprietary 32-bit RISC architecture and a soft CPU core designed by Xilinx for use in their FPGAs. PicoBlaze has a proprietary 8-bit RISC architecture and a CPU core

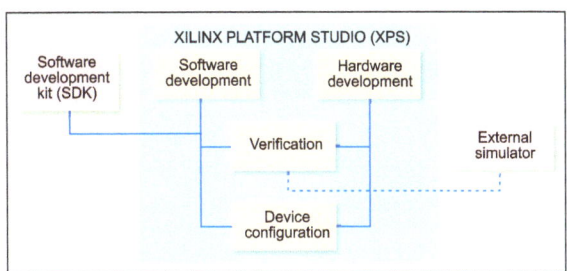

Fig. 1: Embedded-processor design process

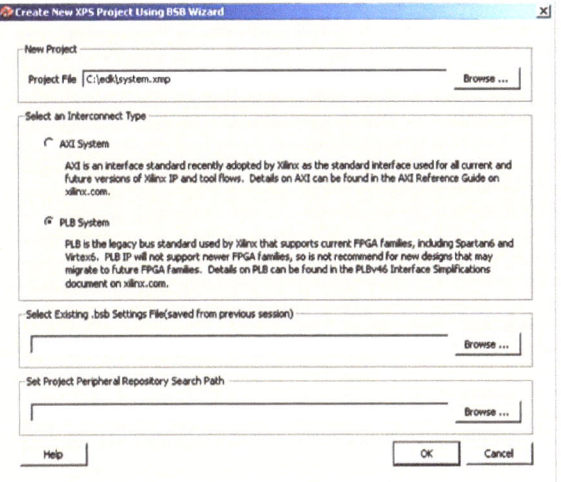

Fig. 2: The path for project file

developed by Xilinx. Nios-II has a proprietary 32-bit RISC architecture and a processor core developed by Altera for use in their FPGAs. It comes in three variants, namely, Nios II/f, Nios II/s and Nios II/e.

The soft processor is typically described in a Hardware Description Language (HDL) or netlist. Unlike the hard processor, a soft processor must be synthesized and fitted into the FPGA fabric. In both hard as well as soft processor systems, the local memory, processor busses, internal peripherals, peripheral controllers and memory controllers must be built from the FPGA's general-purpose logic.

Some of the advantages offered

by an FPGA-embedded processor over a typical embedded processor include customisation, obsolescence mitigation, component and cost reduction and hardware acceleration. The embedded-processor-system designer has the complete flexibility to select any combination of peripherals and controllers. In addition, he can design new peripherals to meet any non-standard peripheral set requirement. As an example, if the designer wants six UARTs for his design, off-the shelf processor with six UARTs is not available. However, the same can be implemented in an FPGA easily.

Some of the disadvantages of FPGA-embedded processor include increased tool complexity and design methodology, which require more attention from the embedded designer. Moreover, unlike an off-the-shelf processor, the hardware platform for the FPGA-embedded processor must be designed. If a standard off-the-shelf processor can do the job, that processor will be less expensive in comparison with the FPGA capable of an equivalent processor design. However, if an FPGA is already used in the system, consuming unused gates or a hard processor in the FGPA will make the embedded-processor system's cost inconsequential.

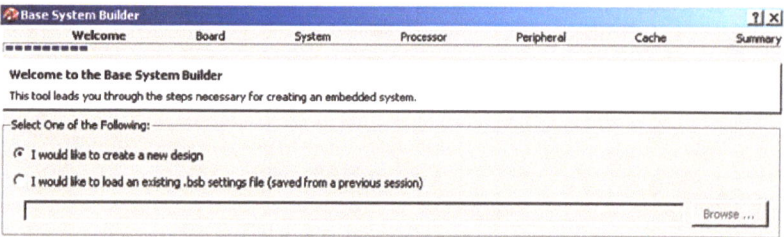

Fig. 3: Base System Builder

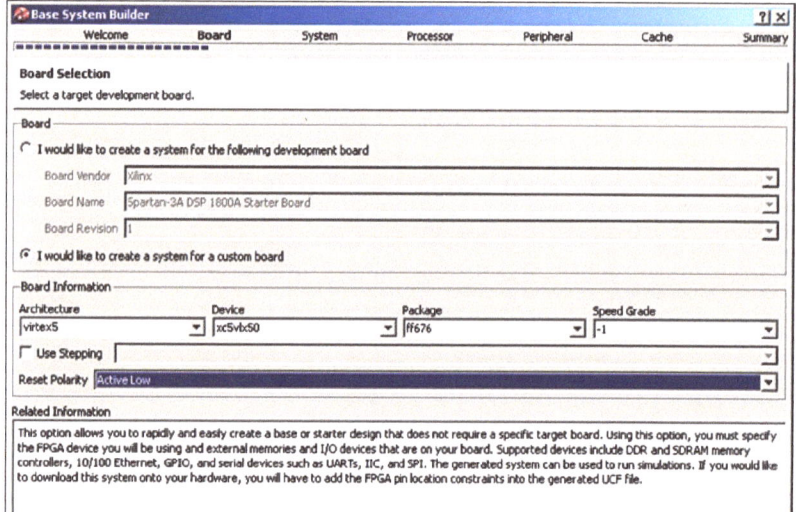

Fig. 4: Custom board selection

To facilitate FPGA-embedded processor design, FPGA manufacturers offer extensive libraries of intellectual property (IP) in the form of peripherals and memory controllers. Some of the peripherals and peripheral controllers provided include general-purpose I/O, UART, timer, debug, SPI, DMA controller, Ethernet and so on. Some of the memory controllers include SRAM, Flash, SDRAM and Compact Flash.

Embedded Development Kit (EDK)

Embedded Development Kit (EDK) is a suite of tools and intellectual property (IP) that allows you to design and implement a complete embedded processor system in a Xilinx FPGA device. The aim of the EDK is to simplify the embedded-processor design process as it provides an umbrella covering all things related to embedded processor and its design. It may be mentioned here that EDK is specific to the FPGAs from Xilinx.

The EDK comprises two main components, namely, the Xilinx Platform Studio (XPS) and the Software Development Kit (SDK). XPS is used primarily for embedded-processor hardware system development, including configuration of the microprocessor, peripherals and the interconnection of these components, along with their respective property assignments. XPS has a base system builder (BSB) wizard that quickly and efficiently establishes a working design. The wizard is generally used to create the foundation of a new embedded design project.

SDK is the recommended software development environment for writing the software code for the embedded processors. Fig. 1 shows the embedded-processor design process.

Creating an EDK project

This section discusses how to create an EDK-driven design task to blink eight LEDs on and off at a frequency of 50Hz.

The different steps for designing the blinking of the LEDs are:

1. Open XPS by selecting Start → Programs → Xilinx ISE Design Suite 13.4_1 → EDK → Xilinx Platform Studio

2. From the dialogue box, select Create New Project Using Base System Builder

3. A window will appear titled Create New XPS Project Using BSB Wizard asking you to specify the folder to place the project. Set the desired path. (We have selected *C:\edk* as shown in Fig. 2.)

4. Then select the interconnect type, either as processor local bus (PLB) or advanced extensible interface (AXI), depending on the FPGA series you are using. As mentioned in the wizard, PLB is a legacy bus standard used by Xilinx that supports current FPGA families, including Spartan 6 and Virtex 6. It will not support new FPGA families. AXI is an interface standard recently adopted by Xilinx as the standard interface used for all current and future versions of Xilinx IP and tool flows. Choose among the two depending upon the FPGA family you are using. We are doing the design using Virtex-5, so we have selected PLB system. Click OK button

5. In the next window select 'I would like to create a new design' option and click Next, as shown in Fig. 3

6. A window appears asking the designer to select the board. If you have a standard development board from Xilinx mentioned in the list, select the first option 'I would like to create a system for the following development board' and fill in the vendor name as Xilinx and your board name. Otherwise, select the second option 'I would like to create a system for a custom board' and fill the board information. The design is being implemented on the ML501 board from Xilinx. As it is not mentioned in the Board Name in the first option, the second option has been selected and the relevant details are filled as shown in Fig. 4. Click Next

7. The next page, that is, the System Configuration page, asks you whether

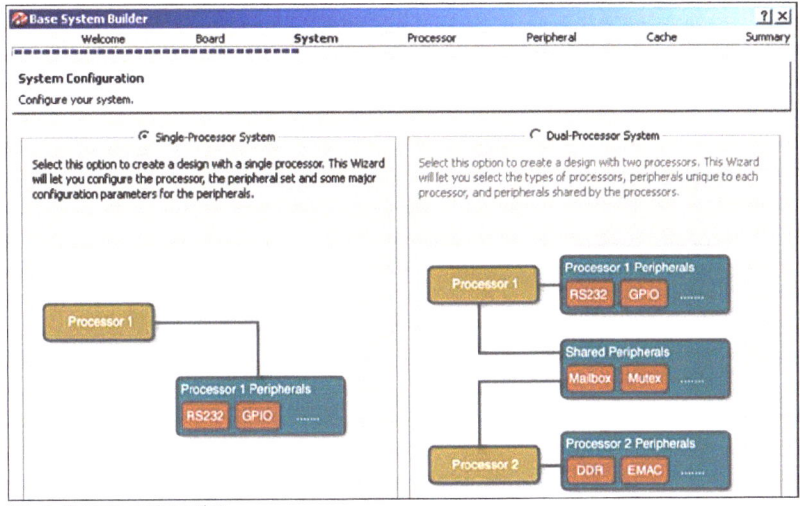

Fig. 5: System configuration

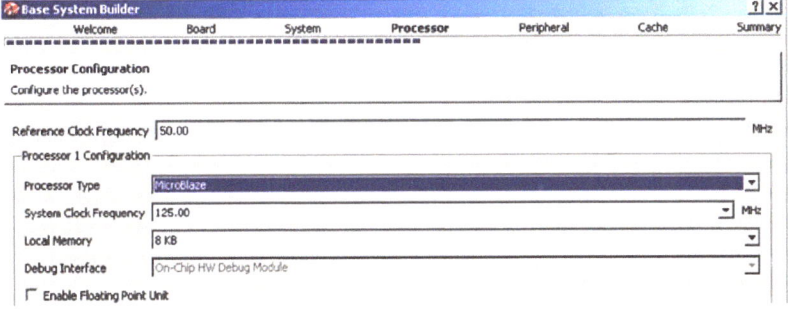

Fig. 6: Processor configuration

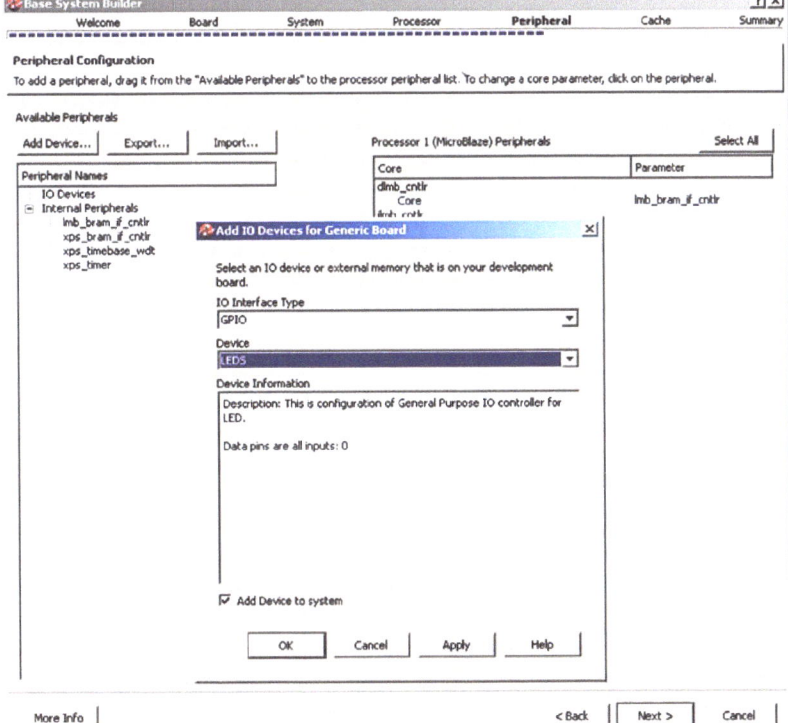

Fig. 7: Peripheral configuration

you want to build a single-processor system or a dual-processor system. Select Single-Processor System and click Next (Fig. 5)

8. The Processor Configuration page appears. Select 50.00MHz as the reference clock frequency. The Virstex-5 FPGAs do not have any hard processor, so Base System Builder will set up a MicroBlaze soft processor. Set the system clock frequency as 125.00MHz and local memory as 8kB, and click Next, as shown in Fig. 6

9. The Peripheral Configuration page appears through which we select the peripherals to put in the design. The peripherals will be connected to the MicroBlaze processor via the PLB, and they allow you to control and access features of the FPGA and external hardware. Click the Add Device button, select the IO Interface Type as GPIO and the device as LEDs, as shown in Fig. 7. Click OK

10. Select the GPIO Data Width as 8, as shown in Fig. 8, and click Next

11. Click Next on the Cache Configuration window

12. The Summary page (refer Fig. 9) gives you a summary of the design created by Base System Builder, showing the PLB memory map, peripherals and the files that will be generated

13. Since we are designing using a custom board, the window shown in Fig. 10 appears, indicating that the user needs to update the user constraints file and set the correct position of the FPGA in the JTAG chain. Click OK

14. The System Assembly View window appears, as shown in Fig. 11. The different regions in the window are explained below for understanding

15. The area marked as (A) contains the project information and has two tabs, namely, Project and IP Catalogue. The Project tab lists the project files like the Microprocessor Hardware Specification (MHS) file, User Constraints File (UCF), Bitgen Options File, iMPACT command file, implementation options file and the ELF file, and the project settings, including the device number, simulation model and netlist type

The MHS file defines the hardware

components, that is, the configuration of the embedded-processor system, and includes the bus architecture, peripherals, processor, system connec-

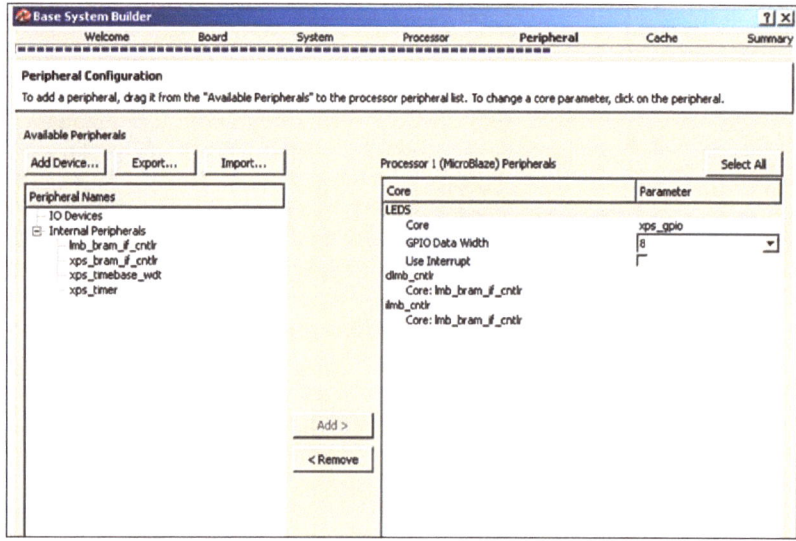

Fig. 8: GPIO data width

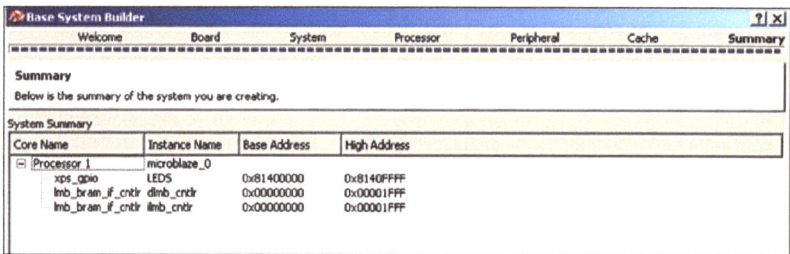

Fig. 9: Summary page

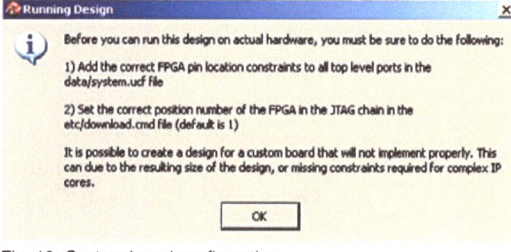

Fig. 10: Custom board configuration

tivity and address space. It serves as an input to the platform generator (platgen) tool. The UCF specifies the timing and placement constraints for the FPGA design. The Bit-gen Options File contains options for the bitstream generation tool. iMPACT command file is a script file used for downloading bit streams using iMPACT. The IP catalogue contains a list of the peripherals or IP cores that the project has access to. This tab is used when you want to instantiate IP cores into the design.

(B) is the Bus-Connectivity area. It shows the interconnections between the different IP cores, memory and the processor. There are two types of buses, namely, the Processor Local Bus (PLB) and the Local Memory Bus (LMB). The vertical line represents a bus and the horizontal line represents a bus interface to an IP core. A hollow connector represents a connection that you can make and a filled connector represents a connection made. To create or disable a connection, click the connector symbol.

(C) is the View-Buttons tab. There are two buttons here, using which you can change between Hierarchical view and the Flat view. Hierarchical view is the default view in the System Assembly View panel in which the information about the design is based on the IP core instances in the hardware platform and organised in an expandable or collapsible tree structure. When you click the directory structure icon, the parts are displayed either hierarchically or in a flattened view.

(D) is the Console Window area. It displays all the textual information, warnings and errors that occur as the changes are made to the project, netlists and bitstreams, etc.

(E) is the System Assembly View in which you can view the project's IP cores and the properties. It displays all hardware platform IP instances using an expandable tree and table format. The IP elements, their ports, properties and parameters, which are configurable in the System Assembly View, are written directly to the MHS file. There

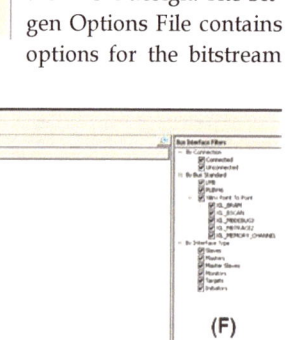

Fig. 11: System Assembly View

Fig. 12: UCF file

are three tabs in the System Assembly View, namely, the Bus Interfaces, Ports and the Addresses. The bus interface

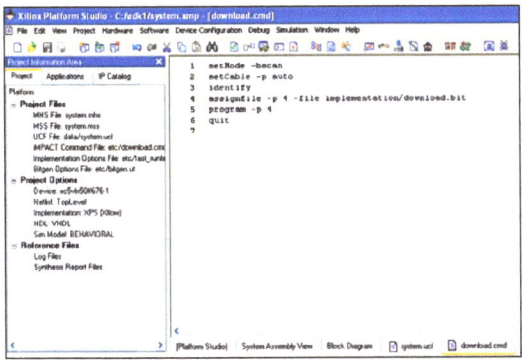

Fig. 13: Download.cmd file

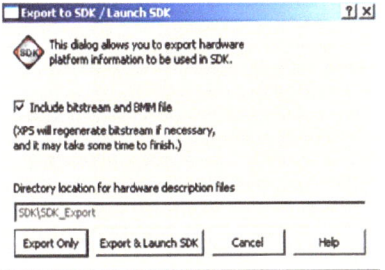

Fig. 14: Export to SDK/Launch SDK

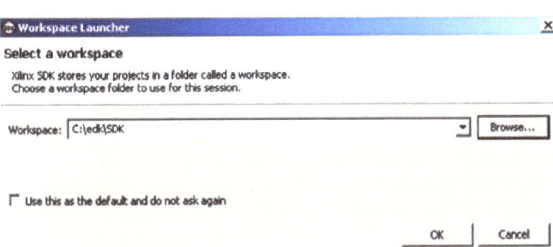

Fig. 15: Workspace Launcher

tab shows the following information for each core:

(i) Instance name (name of the core used to reference the core in the UCF)

(ii) Bus name (the bus to which the core is connected)

(iii) IP type (name of the IP core as in the IP catalogue)

(iv) IP version (the version of the IP core)

The Ports tab shows the user accessible ports of the IP cores and the nets to which they connect. The Addresses tab shows how the IP cores are mapped to the PLBs in the design

(F) is the Filters Pane area. It allows you to filter what is shown in the System Assembly View for simplified view of the design or in the case of many IP cores in the design.

16. Since we are using custom board for designing, the UCF file needs to be updated. Allot the FPGA pins to the LEDs, Clock and RST signals. The UCF file created for the ML501 board is shown in Fig. 12.

17. The download. cmd file is changed, as shown in Fig. 13, as the FPGA is the 4th device in the JTAG chain used. (The num-

ber assigned depends upon the JTAG being used.) Save the project

18. From the XPS menu, select Project → Export Hardware Design to SDK

19. In the dialogue box that appears (Fig. 14), make sure that Include Bitstream and BMM File are ticked and click Export Only

20. The bitstream will be generated and the project will be exported to SDK subsequently. Once it is done, we can progress to the next step

21. Open SDK by selecting Start → Programs → Xilinx ISE Design Suite 13.4_1 → EDK → Xilinx Software Development Kit

22. The first thing that the designer has to do is to enter the SDK workspace. SDK workspace is a folder wherein the software application(s) for a particular EDK hardware design are managed. The folder specified is shown in Fig. 15. You can choose any folder but it is better if the software folder is created inside the main project folder for ease of use

23. SDK opens with a Welcome screen. Select File → New → Xilinx C Project. Next, you have to specify the hardware platform. Click Specify on the window that appears. In the dialogue box that appears, type the name of the project as edk and use the Browse button under the Target Hardware Specification (refer Fig. 16) heading to navigate to C:\edk\SDK\ SDK_Export\hw\system.xml file.

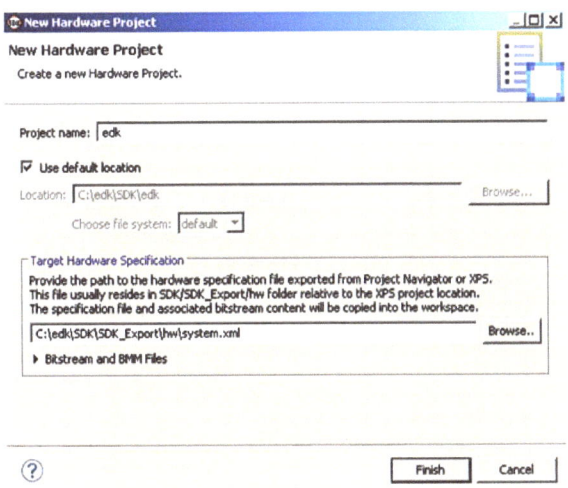

Fig. 16: Target Hardware Specification

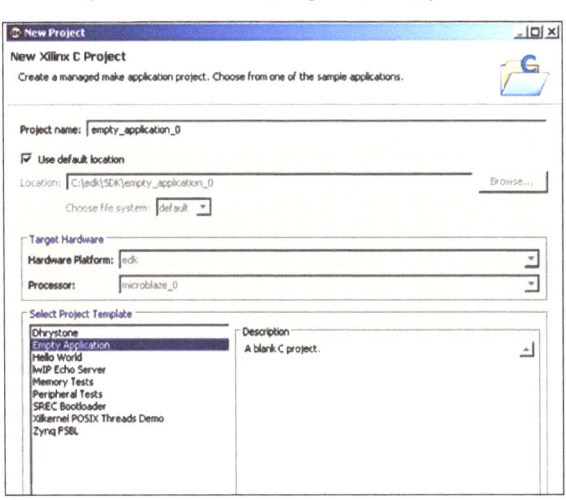

Fig. 17: Select Project Template

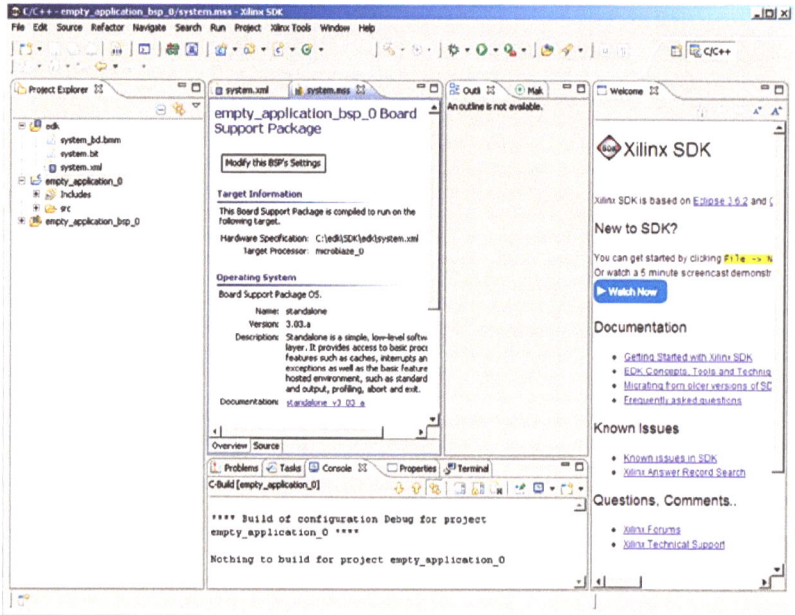

Fig. 18: Xilinx SDK window

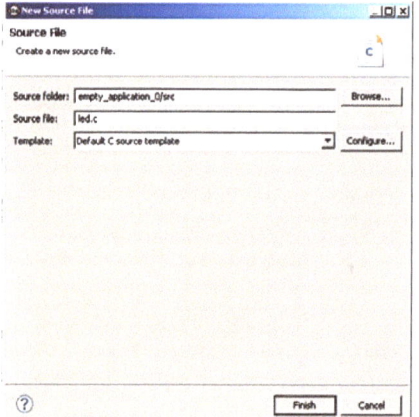

Fig. 19: New Source File window

```
8  #include "xparameters.h"
9  #include "xgpio.h"
10 #include "xutil.h"
11 int main()
12 {
13     unsigned long i ;
14     // init leds
15     XGpio LEDS;
16     XGpio_Initialize (&LEDS, XPAR_LEDS_DEVICE_ID);
17     XGpio_SetDataDirection(&LEDS, 1, 0x00);
18     //out
19     while (1)
20
21     {
22         XGpio_DiscreteWrite (&LEDS, 1, 0xFF);
23         for (i=0; i<0xF0000; i++);
24         XGpio_DiscreteWrite (&LEDS, 1, 0x00);
25         for (i=0; i<0xF0000; i++);
26
27     }
28     return 0;
29 }
30
```

Fig. 20: Program compilation

Click Finish

24. The wizard that follows allows you to create a template software application for the project. The default is 'Hello World' example. We want to create a blank application. Select Empty Application and click Next, as shown in Fig. 17. Click Finish

25. An SDK window appears as shown in Fig. 18

26. In the Project Explorer, open the tree empty_application_0 → src. There is no source file in the folder. We need to add the source file. Select src folder and go to File → New → Source File. In the dialogue box that opens, specify the name of the source file to be created and the folder in which to put it, as shown in Fig. 19. Click Finish. If you now expand the src folder, you will find led.c inside that folder

27. Click on led.c and paste the code given below onto it. Save the file. When you save the file, it is automatically compiled. You should get zero errors, as shown in Fig. 20

The C code first initialises the LEDs port and then sets the port as output port. Then the LEDs are switched on and off using a delay created by the 'for' loop

```
#include "xparameters.h"
#include "xgpio.h"
#include "xutil.h"
//==========Program===========//
int main () {
unsigned long i;
//Init LEDS
XGpio LEDS;
XGpio_Initialize(&LEDS, XPAR_LEDS_
DEVICE_ID);
XGpio_SetDataDirection(&LEDS, 1,
0x00);
//Out
while (1)
{
XGpio_DiscreteWrite(&LEDS, 1,
0xFF);
for (i=0; i<0xF0000; i++);
XGpio_DiscreteWrite(&LEDS, 1,
0x00);
for (i=0; i<0xF0000; i++);
}
return 0;
}
```

Select Xilinx Tools → Program FPGA. The bitstream gets loaded to the FPGA. The LEDs start blinking. If they do not blink, press the reset switch used in the design. On pressing the reset switch they will start blinking.

Conclusion

With this part, the series of articles on Designing with FPGAs is concluded. The articles were written with an aim to provide the electronics engineers, professionals and hobbyists an insight into different aspects of system design using FPGAs. ●

Concluded

□arsha □rawa□ is a s□entist at □aser □□en□□ an□ Te□hno□□□ □enter □□□□□□□a □re□ier □□□□□ □□ wor□in□ in the fie□□□ of □aser□□ase□ □efen□e s□ste□s□□he has □ore than □□ □□ars of □□□□ e□□erien□e in the □esi□n an□ □e□□□□□ent of a □ariet□ of s□ste□s for □efen□e□re□ate□ a□□□□ations□ □he has authore□ two □oo□s an□ □u□□she□ □ore than □□ resear□h □a□ers an□ te□hni□a□ arti□□es

Troubleshooting and Repairing Tips

Do-it-Yourself (DIY) troubleshooting and repairing of electronics gadgets can be fun as it helps you save money and gain useful knowledge. This month we present some websites for the brave hearts who like to troubleshoot and repair electronic devices by themselves

NIRAJ SAHAY

fixya.com

FixYa is a community-based troubleshooting resource that provides consumer-generated, practical tips to help consumers solve problems concerning over 12 million products. Fixya is a place where individuals can share real-world experiences and connect to provide each other practical advice. Today, FixYa continues to empower individuals to repair and improve upon their already-purchased possessions. From fixing cars to cameras and iPhones, FixYans are part of a DIY revolution that helps empower techies, tinkerers and hobbyists across the globe.

http://www.fixya.com/

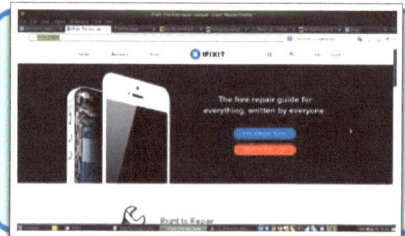

ifixit.com

iFixit is a global community of people helping each other repair things. It was started in 2003 by Luke and Kyle in a dorm room at Cal Poly, San Luis Obispo. It helps thousands of people repair their devices every day. Initially started for Apple products, the website now provides free repair manuals and troubleshooting guides for other products too.

http://www.ifixit.com/

repairfaq.org

Sci.Electronics.Repair frequently asked questions (FAQs) feature Samuel M. Goldwasser's latest and greatest notes on troubleshooting and repairing. The site is text based and there are no unnecessary, superfluous or useless graphics of any kind. The site covers troubleshooting and repair of consumer electronics equipment.

http://www.repairfaq.org/

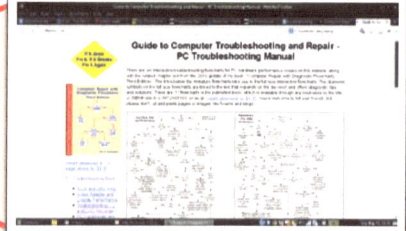

ifitjams.com

When the problem you are trying to fix is not so obvious, it can help to have a flowchart that logically takes you through the process of elimination. The website has details of troubleshooting laptops, PCs and cars based on flowcharts. The site was started by Morris Rosenthal and is a wonderful resource for people who like troubleshooting.

http://www.ifitjams.com/

diy-computer-repair.com

The website helps you discover how to troubleshoot, repair, upgrade and maintain your computer. The purpose of this DIY computer-repair website is to assist you with all the information that you may need to troubleshoot and repair your faulty desktop, laptop or server. The site also hosts blogs to share your experiences on fixing problems.

http://www.diy-computer-repair.com/

Quality Labs and Certification in Demand

The government of India is pushing the manufacturing industry for a huge surge in exports, and this is leading to the growth of the quality labs and certification industry

SNEHA AMBASTHA

A lab that is completely in line with international requirements and is ISO/IEC 17025 certified is known as a quality lab. Laboratories accredited to this international standard demonstrate their technical competency and ability to produce precise test and/or calibration data.

Testing service labs, such as STQC and TUV SUD, are becoming testing arms for the government of India, where they test and release reports that help the government take the decision on registration of products.

The initiatives taken by the government of India, like the Compulsory Registration Scheme (CRS), would force all the electronics goods of different rating, size and variety to be certified at least once every two years.

"Electronics Test and Development Centre (ETDC), Hyderabad addressed a meet with all manufacturers of electronic goods and promulgated the order 'Electronics and Information Technology Goods (Requirements of Compulsory Registration) Order, 2012' which came into effect from April 3, 2013, mandating (testing) compliance of goods to Indian safety standards and compulsory registration with Bureau of Indian Standards (BIS) by the stakeholders, including manufacturers or importers or stockists or distributors," says P. Chow Reddy, managing director, Interleaved Technologies.

Overall growth factors

The growth of these labs is directly dependent on the requirements of the manufacturing industry.

Export of goods. The government of India is pushing the manufacturing industry for a huge surge in exports,

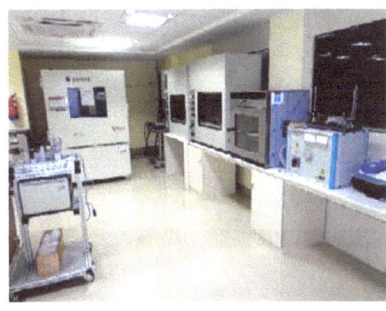

STRENGTHS
• The global information
• The know-how of the products
• The product requirements
• The regulatory requirements of various countries

OPPORTUNITIES
• Growth of the Indian Electronics Industry
• Government subsidiaries and investments
• Tax benefits on technology growth
• Improvement of India's share in domestic manufacturing supply

SWOT

WEAKNESSES
• High dependence of the labs on local testing and requirements
• Less focus on export oriented testing and certification

THREATS
• Lot of price erosion in terms of services
• Thrilled manpower

Fig. 1: SWOT analysis for quality labs and certification

and this is leading to the growth of the quality labs and certification industry. The question is how? All electrical and electronics goods that are to be exported to other countries have to fulfil certain safety norms and certification requirements. The World Trade Organisation (WTO) and Technical Barrier to Trade (TBT) recommend the adoption of these international standards and ensure that they are certified for the same in order to facilitate their trade. Thus the increase in exports is increasing the demand of compliance measures, thereby increasing the demand for quality test labs and certifications.

Suresh Kumar, senior vice president, consumer product services, TUV SUD says, "Today, the Indian market is competing with the global expertise on the electronics products where we form a very small part exporting less than one per cent of that of the global exports. Then, there are other factors, like the products manufactured here

meet very little safety criteria as compared to the products from the US or Europe. So the main purpose of labs in India is to assist the manufacturers to meet the safety criteria for their products by transferring the know-how of what they gain globally."

Regional impacts. The productivity of labs in a country is not only dependent on the type of industry, it also has certain regional impact. The policy and regulations for products in any particular region depend on both the consumers and their stakeholders, such as buyers, industrial houses, retail markets and other entities that help maintain the minimum quality level of these products and goods.

Each region has its own electrical parameter and environmental conditions which increase the demand for labs with those testing and certification abilities. This helps small regional labs grow, although this growth is not to a large extent.

New and upcoming products. Currently, not only in India but all across the world we can see a huge demand for new products with latest technologies. Even the government of India is encouraging new Wi-Fi products along with photovoltaics. Government has also bifurcated the schemes related to electronics into different subsections to encourage the production of new products.

Now, as per the schemes, there are new standards that involve new testing tools, testing criteria, new skills and new certification requirements at times. Increased investment by the government in the manufacturing sec-

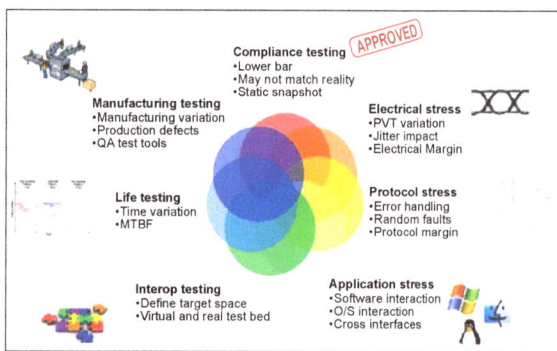

Fig. 2: Different testing approaches by the quality labs
(Courtesy: Granite River Labs)

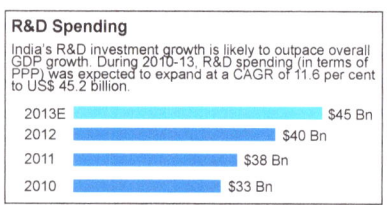

R&D Spending

India's R&D investment growth is likely to outpace overall GDP growth. During 2010-13, R&D spending (in terms of PPP) was expected to expand at a CAGR of 11.6 per cent to US$ 45.2 billion.

2013E		$45 Bn
2012	$40 Bn	
2011	$38 Bn	
2010	$33 Bn	

Fig. 3: Expenditures on research and testing
(Courtesy: India Brand Equity Foundation (IBEF)

tion further adds to production, thus paving a way for the requirement of good, high-end quality labs for testing.

Increasing demand of certification. If we keep the different aspects aside and look at the certification industry itself, we find that one of the major growth factors for the quality labs is the enforcement of certification. The different schemes by the government, the quality requirements by the consumers and even the industries themselves force manufacturers to get their products certified in order to ensure that they comply with the necessary basic standards. Ankan Mitra, vice-president, SMTA India Chapter explains, "The way this product certification industry works is that, if a company wants to come for manufacturing in India, they would want to

Although the growth of the labs in India is slow currently, plenty of factors have come up in the recent years to bring India at par with the other nations

In one of his interviews, Kapil Sibal, Union Minister of Communications and Information Technology said that, with expansion of 3G and 4G/LTE, the requirement of testing labs in India has increased and the ministry would allow self-certification process going forward. He also added that India, with the manufacturing base market of 1.2 billion individuals, needs testing facility within the country

manufacture and then get it certified here itself, before taking it back to West or before selling it here in India. This is as opposed to manufacturing it in West and then getting it certified here. So taking the whole South East Asia into consideration, certification enforcement acts as the driving force for this industry."

Growth of other related industries. Growth of the electronics industry, IT and software industry, and increasing consumer demands, are all leading to the growth of the quality testing labs and universal certification requirements. If we take the example of mobile phones in India, in 2003 this number was about 18 million, which increased to about 853 million in 2014 and is now expected to grow more than double in the coming years. With this type of growth in the consumer segment, their quality and safety requirements are also increasing.

As per a report by Indian Brand Equity Foundation (IBEF), the labs that have been set up in India and authorised by India's Standardisation Testing & Quality Certification Directorate (STQC) are supposed to be well equipped to certify the IT systems that are supplied by local and global vendors as safe to connect to the country's core infrastructure networks. They also reported that the software companies, such as Tech Mahindra and Wipro, have already evinced an interest to set up test labs in India.

Government policies and their effects

As per a newsletter by the government of India, Department of Electronics

and Information Technology (DeitY) and Ministry of Communications and Information Technology, they have come up with a reimbursement scheme for testing and certification required to export electronics goods. The scheme is to support and recognise the micro, small and medium-scale enterprises in the electronic system design and manufacturing (ESDM) sector. Under the scheme, the grant in aid is ₹ 20 million for 800 models (max) while it is ₹ 120,000 (max) for one model.

The National Policy on Electronics aims to create an institutional mechanism for developing and mandating standards and certification for electronic products and services to strengthen nation-wide quality assessment infrastructure.

Although these government policies with respect to quality labs do not affect Indian industry, these help the R&D sector grow. As per Reddy, "UL recently started tests in UL 1703 standards, thus facilitating Indian solar PV panels to test and certify."

IBEF reported that, with respect to the engineering R&D, India is one of the leading offshore destinations in delivering these services with a market share of 22 per cent. The market in India is expected to grow at a compounded annual growth rate of 14 per cent, from $14.7 billion in FY12 to $42 billion by 2020, and is also expected to outpace the information technology growth rate in India.

Challenges in India

To our amazement, while there are plenty of growth factors available, we also see a major threat to this industry, as the labs here are not always well versed with the requirement of test facilities and capabilities.

Lack of test facilities. Considering the manufacturing industry for the domestic, industrial and defence electronics products specially, there are no special test facilities or tools. This leads to the wastage of a large amount of time searching, knowing and getting well equipped with the resources needed. Above all, the labs in India are also short of the latest tools available as

Forecast

Suresh Kumar says, "Most regulations come through governmental organisations globally. However, there are certain segments like automotive, which have a consortium of expertise that bring forth some special standards. For example, Forbes, Clintwear and GM joined together and have come up with some standards for the automotive industry but then restricting myself to the consumer market, the products have to comply with the local government standards or international standards."

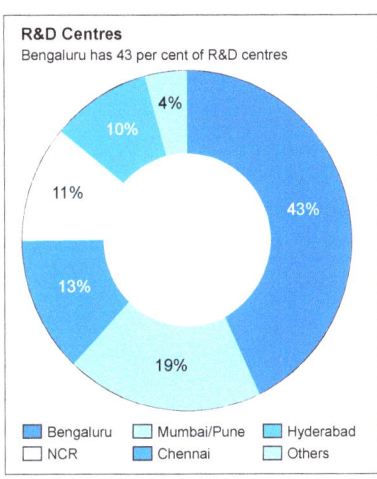

R&D Centres
Bengaluru has 43 per cent of R&D centres

- 43% Bengaluru
- 19%
- 13%
- 11%
- 10%
- 4%

☐ Bengaluru ☐ Mumbai/Pune ☐ Hyderabad
☐ NCR ☐ Chennai ☐ Others

Fig. 4: Research and development centres in India (Courtsey: India Brand Equity Foundation (IBEF)

compared to the labs abroad.

Lack of skilled manpower. Skilled manpower is somewhat related to the test facilities, as the skills are also required with the usage of the tools and facilities rather than just the implementation of the procedures. Vamshi Kandalla, global GM and VP, Granite River Labs says, "It is but obvious that an individual with strong knowledge in the certifications and test facilities in the country can operate as a consultant to a number of manufacturing industries and development centres."

No supportive policy for setting up labs. Although the government has come up with schemes to build a few labs in India, there is no conducive policy for someone setting up a testing lab. Peeyush Gupta, director, sales and marketing, UL says, "In India, setting up a laboratory involves high costs and advanced infrastructure development. In some cases, such as setting up

"Overall impact of the government schemes and policies is not affecting the Indian industry but the R&D section of the industry" —P. Chow Reddy

a top-quality mobile testing laboratory or a wireless testing laboratory, the cost varies between 200 million and a billion rupees. Unfortunately, there are no existing regulatory policies or government schemes that support infrastructure development or incentivise quality facilities to address specific market regulations." Sometimes, it is difficult to set up a lab as the kind of infrastructure required for this is so immense.

From tens of millions of rupees, some can require ₹ 500 million investment. One is required to spend anywhere from ₹ 2000 million to ₹ 5000 million to set up a good-quality mobile testing lab or a wireless testing lab.

Lack of public private relationship. There are some private labs in India that are working in the field of testing since over 100 years in a global environment. They have a lot of knowledge and expertise for not only testing but also for standard-to-standard citing. Generally, the Indian government looks at these private labs only for innovation testing or infrastructure.

The government should open up in terms of policies and standards for private players. There is a need of firm commitment in terms of regulations,

Challenges to the industry are not few, but a better understanding between the government of India and the private testing labs can help eliminate the major challenges

"UL is committed to working with manufacturers, industry associations and government bodies to accelerate adoption of safety standards and certification procedures. Our aim is to ensure that the consumer has access to high-quality products as a result of regulatory testing and standardisation of products and services." —Peeyush Gupta

Labs to come up

Government of India has planned to set up some labs in India to support the growth of the electronic system design and manufacturing (ESDM) industry at a CAGR of 9.9 per cent. In view of this, the Department of Electronics and Information Technology (DeitY) has been proposed to invest ₹ 650,000 million in the area of manufacturing of electronics and semiconductors. Thus the first lab on semiconductor characterisation in India is expected to be set up in Bengaluru in the next six months, the success of which would lead to the set up of another lab in Bhubaneswar that is expected to be launched in next one year.

A report on Karnataka ESDM policy suggests that the government has decided to set up three ESDM innovation centres within Karnataka. The main idea behind these centres is to provide facilities for testing and development, to have characterisation labs, and compliance and certifications labs. The first amongst these would be housed at IIIT Bengaluru.

Another report by India Electronics and Semiconductor Association (IESA) suggests that BIS is to set up a test lab in Khurda cluster to help the preferential access to the products that are manufactured in India.

MAJOR CONTRIBUTORS TO THIS REPORT

Ankan Mitra,
vice-president,
SMTA India
Chapter

C.S. Bisht,
scientist 'G' and
senior director,
IT Centre, STQC
Delhi

**P. Chow
Reddy,**
managing director,
Interleaved
Technologies

Peeyush Gupta,
director, sales
and marketing,
UL

Suresh Kumar,
senior vice
president,
consumer product
services, TUV SUD

**Vamshi
Kandalla,**
global GM and VP,
Granite River Labs

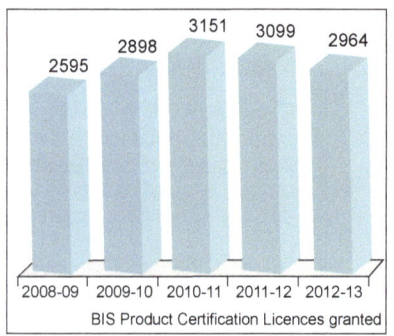

Fig. 5: Product certificate licences (Courtsey: BIS)

policies and incentives from the government. Otherwise, it becomes very difficult for the industry to come forward and take the initiative or to make the first move.

Consider a situation where there is a small change in standards by BIS and the private test labs have not been informed about it in time. They may have tested 500 products by the time the new standard is announced. Now what should the test labs do with the 500 test reports? There would be extra time and manpower required to test according to the new standards for the correct certification.

Gupta explains, "Over the last few months there have been several requests pertaining to certified components, power supply and power cords. These requirements have not been restricted to just UL but were also applicable to other private, ERTL or government labs, with instances where reports have had to be withdrawn and revised accordingly."

Solutions implemented

The manufacturing sector is taking steps to see that they adopt sustainable practices. The labs now ensure that their services are not limited to testing but they also support the manufacturers for proper systems implementation. The labs are also ensuring that the manufacturers follow proper social compliance standards and maintain certain occupational safety in their work premises.

As per a report by BIS, they are taking certain basic measures to ensure the growth of the labs and the certification industry, which includes training and the labs' recognition

schemes. There are regular trainings for the manpower working in the labs to ensure their awareness of latest developments on international platforms. BIS is training officers from its own labs on international standards like IS/ISO/IEC 17025 and is also providing summer training to the students from different universities and colleges on the laboratory-related works.

BIS has also come up with the Laboratory Recognition Scheme (LRS) that helps the outside laboratories (OSLs) to get recognition. These OSLs are the ones that help labs like BIS with testing facilities during situations such as a large number of samples for testing, or BIS testing equipment being out of order. In order to support this scheme, about seven OSLs have been recognised in the year 2012-13 to test products covered under electronics and information technology goods, such as IT (IS 13252), AV (IS 616) and microwave ovens (IS 302-2-25).

C.S. Bisht, scientist 'G' and senior director, IT Centre, STQC Delhi says, "STQC has come up with the scheme for approval of IT test laboratories (ITTLs) for software solutions. STQC will be the nodal agency managing this approval scheme. The scheme requires ITTLs to implement a domain-specific laboratory 'Quality Management System' based on ISO/IEC 17025 standard. The IT test laboratory will be assessed as per ISO/IEC 17025 general requirements for the competence of testing and calibration laboratories. ITTLs will also be assessed as per ISO/IEC 25051 (ISO/IEC 29119 under development) for test management activities. STQC approval body (ITTL) will operate as per ISO/IEC 17011: 2004 standard 'Conformity Assessment-General requirement for accreditation bodies accrediting conformity assessment bodies'."

Vamshi adds, "There are a number of OSLs in India presently that cannot be compared to the giant service and solution providers, though they have their own value and their existence here is important as per the requirements." •

The author is a technical journalist at EFY

Indian UPS market to grow at 11 per cent CAGR

In its recent India UPS Market Outlook 2018 report, industry research and consultancy firm RNCOS has revealed that, owing to the significant demand for uninterrupted power supply, the Indian UPS market is expected to grow at a healthy rate of 11 per cent from 2014 to 2018.

The UPS market in India has been predominantly dependent on low-range UPS systems (up to 25kVA) constituting over half of the total UPS market. However, with the growing utilisation of UPS systems in larger industries, the trend is more likely to change and contribute to its growth.

The report reveals that new technologies are gaining greater penetration into the UPS market in India, and there's a gradual inclination towards solar-powered UPS systems. The growth is also fuelled by an increase in demand from new markets, primarily in tier II and III cities.

Biometrics market could grow at 44 per cent

Research and Markets in its recent Biometrics Market in India 2014-2018 report has forecast that the Indian biometrics market is likely to grow at a CAGR of 44.22 per cent over the period 2013-2018. The report is based on an in-depth market analysis in the country with inputs from industry experts.

The report reveals that, while fierce

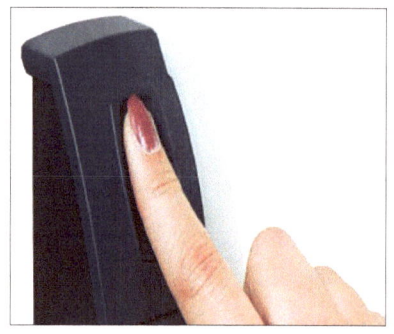

In Focus

R.S. Sharma joins as new DeitY secretary

Ram Sevak Shama, a Jharkhand cadre IAS officer, has joined the Department of Electronics and Information Technology (DeitY) as secretary. He succeeds J. Satyanarayana who retired from the post on 30th April. Sharma was earlier director general and mission director of the Unique Identification Authority of India (UIDIA) where he played a vital role in developing Aadhaar card concept along with former UIDIA Chairman Nandan Nilekani.

Vidyashankar is the new IESA president

After retiring as principal secretary, Commerce & Industries Department, Government of Karnataka, M.N. Vidyashankar, IAS has joined India Electronics and Semiconductor Association as its president. He succeeds P.V.G. Menon.

Nisha Gupta appointed Honeywell Automation India director

Honeywell Automation India Ltd (HAIL), the Indian subsidiary of US-based Honeywell Inc., has appointed Nisha Gupta as its new director. Nisha now becomes the first woman director ever in HAIL's history. Honeywell Automation India is a leading integrated automation and software solutions provider in the country.

Samsung India chief Vineet Taneja resigns

Vineet Taneja has reportedly resigned from his post at Samsung India. He was the country head for the mobile and digital imaging arm of the company. He had joined the South Korean electronics giant from Bharti Airtel in 2013.

Amritanshu Khaitan appointed as Eveready Industries India MD

Eveready Industries India has appointed Amritanshu Khaitan as its new managing director. The company, which makes batteries, torches and lighting instruments, reported a net profit of ₹ 136 million for the financial year ended 31st March, up from ₹ 50.8 million reported in the previous year.

competition from inexpensive non-biometric technologies posed a major deterrent to growth, the increasing number of government-initiated biometrics projects was breathing fresh life into its growth prospects. Further, the biometrics market in India has also been witnessing the gradual integration of biometrics in smart cards, another driving force in its growth story.

Solar and LED markets likely to flourish under the new PM

Even as an estimated 300 million plus people still remain unconnected to the grid in India, with the power demand here likely to double by 2020, the country sees fresh hope in Prime Minister Narendra Modi. Modi has made his intentions already clear: he wants every home in the country to have at least one light bulb and have solar energy to

power the same, by 2019. This provides a massive opportunity of sorts for solar and LED companies.

On an average, nearly 27 per cent of the country's power succumbs to theft and distribution losses. Further, massive blackouts reduce its GDP by nearly 1.5 per cent annually, quotes the World Bank. Amid all the loss, the power that reaches the end user is a massive burden on his/her pockets.

Modi has not been very vocal on actually bringing about an LED revolution in India. However, considering the fact that LEDs consume far less power than traditional lighting solutions, and can be further coupled with networking to save even more of it, LED is clearly the way to go! Modi brought about a successful solar program in the state of Gujarat under his tenure as chief minister there. The program has

Calendar of Forthcoming Electronics Fairs/Exhibitions/Seminars/Events

Name, Date and Venue	Topics to be covered	Contact address for details
Communic Asia 2014 June 17-20, 2014 Basement 2, Level 1 and 3, Marina Bay Sands, Singapore	An event that covers the entire ICT ecosystem from 4G/LTE, AR and innovations, content security management, FTTx, mobile apps, mobile broadband, mobile devices, RF and cables, telecom energy and power system and many others	Namrata Bansal/Ritu Rupramka, Mileage Communications India Pvt Ltd 119, Sishan House, 2nd Floor, Shahpur Jat New Delhi 110049 Phone: +91-9013442403, +91-9650927416 E-mail: namrata@mileage.in, ritu@mileage.in
Elasia 2014 July 25-28, 2014 Pragati Maidan, New Delhi	An international exhibition on power, electrical and lighting	Manish, Triune Exhibitors Pvt Ltd #25, 3rd Floor, 8th Main Road, Vasanthnagar, Bengaluru 560052 Phone: +91-9310492923, +91-80-43307474/ +91-80-22352770, E-mail: info@triuneexhibitors.com
Defence and Aerospace SES 2014 **5th Strategic Electronics Summit** July 30-31, 2014 Bangalore International Exhibition Centre	Conferences on defence requirements, exhibition showcasing Indian defence electronics sector, buyer-seller meetings and industry study on strategic electronics sector opportunities	Rajesh Rawat, ELCINA Electronic Industries Association of India ELCINA House, 422 Okhla Industrial Estate, Phase III, New Delhi 110020 Phone: (011) 26924597, 26928053; +91-9911445890 E-mail: info@elcina.com; rajesh@elcina.com Web: www.elcina.com
IPCA-EFY Expo 2014 August 6-8, 2014 Auto Cluster Exhibition Centre Pune	Provides opportunities for all OEM manufacturers, LED, solar, automobile, medical, defence and R&D material suppliers and end-users to interact with each other	Indian Printed Circuit Association #2711, 2nd Main, HAL 3rd Stage New Thippasandra, Bengaluru 560075 Phone: (080) 25210109, 25210309 E-mail: ipca@ipcaindia.org
NEPCON South China 2014 August 26-28, 2014 Shenzhen Convention & Exhibition Center, China	Sourcing platforms for South China's electronics manufacturing industry	Tim Wang Phone: +86-21-2231-7016 E-mail: tim.wang@reedexpo.com.cn Web: http://www.nepconchina.com/ehome/
IFA 2014 September 5-10, 2014 IFA Exhibition Grounds, Berlin	A global trade show for consumer electronics and home appliances, presents the latest products and innovations	Nicole von der Ropp, Press Officer Messe Berlin GmbH, Messedamm 22 14055 Berlin/Germany Phone: +49-30-3038-2217 E-mail: vonderropp@messe-berlin.de Web: http://b2b.ifa-berlin.com/
ELCINA-EFY Awards September 12, 2014 India Habitat Centre, New Delhi	Awards for excellence in electronics hardware manufacturing and services	Arun Singh, EFY Group D-88/3, Okhla Industrial Area, Phase 1 New Delhi 110020 Phone: 26810601/2/3, +91-8800094213 E-mail: efyexpo@efyindia.com
Electronica India 2014 and Productronica India 2014 September 23-25, 2014 BIEC, Bengaluru	Covers the whole spectrum of the electronics industry from electronics production to electronic components	Kavita Chhatani, Project Manager, MMI India Pvt Ltd Phone: +91-9819418496; Fax: +91-22-42554719 E-mail: kavita.chhatani@mmi-india.in Web: www.electronica-productronica-india.com
3rd Electronics Rocks 2014 October 10-11, 2014 Nimhans Convention & Exhibition Centre, Bengaluru	A platform for design engineers, R&D engineers, entrepreneurs, academicians, hackers and hobbyists including talks, workshops, discussions, product launches and design challenges	EFY Enterprises Pvt Ltd D-87/1, Okhla Industrial Area, Phase 1 New Delhi 110020 Phone: 26810601/2/3, +91-8800094213 E-mail: electronicsrocks@efyindia.com Web: www.electronicsrocks.com/
OSI Days 2014 November 7-8, 2014 NIMHANS Convention Center Bengaluru	Open Source conference in Asia that aims to nurture and promote the open source ecosystem in the sub-continent	EFY Enterprises Pvt Ltd D-87/1, Okhla Industrial Area, Phase 1 New Delhi 110020, Phone: +91-088000 94211 E-mail: atul.goel@efyindia.com
Electronica/Productronica 2014 November 11-14, 2014 Munich, Germany	Electronic components, production equipment, systems and applications	MMI India Pvt Ltd, Mumbai Phone: (022) 42554700, 42554723 E-mail: Andrea.dsouza@mmi-india.in
Intersolar India November 18-20, 2014 Bombay Exhibition Centre, Mumbai	Exhibition and conference for the solar industry featuring photovoltaics, PV production technologies, energy storage and solar thermal technologies	Brijesh Nair, MMI India Pvt Ltd 507 & 508, 5th Floor, Cardinal Gracias Road Opp. P&G building, Chakala, Andheri (E) Mumbai 400099, Phone: (022) 42554700 E-mail: brijesh.nair@mmi-india.in
2nd EFY Expo- Western India Edition 2014 November 26-28, 2014 Bombay Convention & Exhibition Centre, Mumbai	An event for manufacturers, engineers and traders to source electronics components, products and services and to find latest offerings and dealers and distributors for your products in the western electronics industry	EFY Enterprises Pvt Ltd D-87/1, Okhla Industrial Area, Phase 1 New Delhi 110020 Phone: 26810601/2/3, +91-8800094213 E-mail: efyexpo@efyindia.com Web: www.west.efyexpo.com

already created nearly 900MW of solar capacity in the ground.

Mid IR sensors market worth US$ 7 billion by 2019

Mid IR sensor market at $789 million in 2012 could reach US$ 7 billion by 2019 as price performance increases and unit costs decrease from $ 3000 per unit to $ 300 and even to $ 8 or less per unit, according to ReportsnReports.com. The decrease in size of units from bench-top size devices to portable units makes them more useful across the board in every industry.

Applications anticipated to gain market traction include spectroscopic and bio-medical imaging, materials characterisation, standoff explosive detection, microscopy and non-destructive testing. Spectroscopy and imaging measurements are now easier, faster and more cost-effective than ever before. Homeland security, military communications, infrared countermeasures, chemical-warfare agent detection, explosives detection, medical diagnostics, imaging and industrial process controls, fire detection and remote gas-leak detection, pollution monitoring and real-time combustion controls are among the uses for the mid IR sensors.

Wearable tech prices dropping

Wearable tech is becoming more affordable now than it has ever been, thanks to Chinese wearable device makers, according to China wholesale company Chinavasion. Interest in smart garments and wearable tech has exploded over the last twelve months with tech companies of all countries producing smart watches, watch phones, video glasses and more.

Currently, wearable devices are on the very bleeding edge of cool technology. So most come with an equally high price tag, with gadget makers trading on both the novelty of the products and the company's branding. Smart watches, which only reach their full potential when paired with a smart phone, cost anywhere between 150 and 300 US dollars.

While western electronics manufacturers attempt to justify 1000 dollar prices with gimmicks like 3D capabilities

Calendar of Forthcoming Electronics Fairs/Exhibitions/Seminars/Events

Name, Date and Venue	Topics to be covered	Contact address for details
Electronics For You Expo 2015 February 26-28, 2015 Hall 7 (A, B, C, D, E, F, G, H) Pragati Maidan New Delhi	An expo covering the complete electronics ecosystem, including innovation, manufacturing, product design and product sales	EFY Enterprises Pvt Ltd D-87/1, Okhla Industrial Area, Phase 1, New Delhi 110020, Fax: (011) 26817563 Phone: 26810601/2/3, E-mail: efyenq@efyindia.com Web: www.efyexpo.com
EFY Awards March 13, 2015 Bengaluru	An attempt to give recognition to the leading enterprises and individuals in the Indian electronics field	EFY Enterprises Pvt Ltd D-87/1, Okhla Industrial Area, Phase 1 New Delhi 110020 Phone: 26810601/2/3, +91-8800094213 E-mail: efyawards@efyindia.com Web: www.efyawards.com

Look up under 'Events' section in www.electronicsforu.com for a comprehensive list

Since this information is subject to change, all those interested are advised to ascertain the details from the organisers before making any commitment.

Snippets

Internet of Things market growing substantially

According to a new market research report, the value of Internet of Things (IoT) market was worth $1029.5 billion in 2013, and is expected to reach $1423.09 billion by 2020, at an estimated CAGR of 4.08% from 2014 to 2020. Some of the critical technology trends that are expected to have a huge impact on IoT evolution are IPV6, sensor proliferation, cloud computing, big data and faster communication standards such as 4G-LTE and beyond.

US, China, Taiwan dumping solar products in India

India has reportedly found solar products sourced from the US, China and Taiwan dumped in the local market. According to a Bloomberg News report, such dumping has caused "material injury" to domestic manufacturers like Indosolar, Websol Energy System and Jupiter Solar Power. According to India's Ministry of Commerce & Industry, over 20 companies have sold in the country at a price which is less than half of the regular price in their home markets.

Micromax now manufacturing smartphones in India

Micromax has started manufacturing smartphone handsets as well as tablets at its Rudraprayag facility in Uttrakhand, though the production is at its initial stage. According to some reports, Micromax is the second largest player in India for both smartphones and tablets with 13 and 16 per cent market share, respectively.

Samsung, Qualcomm invest in Indian chip venture

India's wearable devices chip start-up, Ineda Systems, has received a major investment of ₹ 1020 million from Samsung and Qualcomm. The company is developing a wearable processor with an always-on battery that can last a whole month. The aim is to supply chips for the multitude of devices that are coming with the new age of wearable computing and the Internet of Things.

Su-Kam to invest in solar rooftop systems

Power back-up solutions provider, Su-Kam Power Systems is all set to make an investment of ₹ 600 million for installing solar rooftop systems on as many as 27,000 houses in Tamil Nadu. It has already invested around ₹ 300 million in setting up solar systems across 11,400 houses as of now.

Firm plans solar roads to power entire America

Solar Roadways in the USA wants to turn American roads into giant energy farms. It is seeking a million dollar in funds on crowd-funding website Indiegogo to pave the highways with thick LED-lit glass solar panels, whose modular system can withstand the heaviest of trucks. These panels can be installed on roads, parking lots, driveways, sidewalks, bike paths, playgrounds and literally any surface under the Sun. A nationwide system could produce more clean energy than the country uses as a whole.

or massive virtual screens, China makers are able to provide these features with prices under 300 US dollars. Though wearable gadgets have just started to capture the world's attention, they are essentially using technology that's tried, tested and not costly to produce.

Cloud-based home-management market set for rapid growth

Cloud-based home-management systems that allow users to remotely control household features like lighting and air-conditioning are set for rapid growth in the coming years, with the installed base set to rise by a factor of eight from 2013 to 2018.

The global installed base of cloud-based home-management services is projected to grow to 44.6 million at the end of 2018, up from 5.6 million at the end of 2013, according to IHS Technology. The installed base this year will surge 63 per cent to 9.1 million.

Smart-home specialists are expected to play a major role in the deployment of smart home services globally. This category of companies includes both home automation providers that are moving toward a cloud-based model, alongside a new breed of connected home specialists, such as Nest, Revolv and SmartThings.

BSNL comes up with its own e-mail service

BSNL has joined hands with Jaipur-based Data Infosys to develop BSNL Xgen personal and enterprise email service. The e-mail will reportedly be free for the broadband users all across the country. Others will have to pay for it.

With this service, users can schedule the SMS and e-mails in advance, which will be delivered at the desired time. The messages scheduled to be delivered at a later moment will be stored on the servers. The service will allow the storage of files up to one GB to be attached/sent. Users will be able to transmit as many as 5000 e-mails on a daily basis.

BSNL Xgen email service comes with the OTP (one-time password) feature, which helps in additional protection of the account and also allows the user not to remember the password every time he logs in.

Check efytimes.com for more news, daily

COMPONENTS

Nanopower sensor ICs

Digi-Key Corporation has just announced stock of Honeywell's new Nanopower line of magnetoresistive sensor ICs. The ICs respond to either a North or South pole applied in a direction parallel to the sensor.

The ICs use a very-low average current consumption and a push-pull output, which does not require a pull-up resistor, operating from a supply voltage as low as 1.65V. The SM351LT series is for applications requiring ultra-high magnetic sensitivity and a very-low current draw (360nA typical) while SM353LT series is for applications requiring very high magnetic sensitivity and a very-low current draw (310nA typical).

The devices are supplied in the sub-miniature SOT-23 surface mount package on tape and reel for use in automated pick-and-place component installation. The ICs reduce system cost and provide a durable alternative to reed switches in low-power battery applications, including industrial, medical, white goods and consumer electronics.

Digi-Key Corporation
Phone: 000-800-100-1274
Website: www.digikey.in

Low-voltage power MOSFETs

ON Semiconductor has introduced a family of six N-channel metal oxide semiconductor field effect transistors (MOSFETs).

The new NTMFS4Hxxx and NTTFS4Hxxx series of MOSFETs are ideally suited as switching devices for a wide range of applications including server and networking equipment and high-power-density DC-DC converters, or to support synchronous rectification in point-of-load (PoL) modules. Versions of these MOSFETs are available with or without an integrated Schottky diode that can help engineers achieve even greater efficiency.

Best-in-class RDSon performance of 0.7 milliohms and low input capacitance of 3780pF ensure conduction, switching and driver losses are minimised. The MOSFETs offer improved thermal performance and low package resistance and inductance compared to existing devices.

Om Semiconductor Technology India Pvt Ltd, Bengaluru
Phone: (080) 66604800, 22246452
Website: www.onsemi.com

Smart-Meter ICs

STMicroelectronics has introduced new smart-meter chips that enable utility companies to improve billing by measuring accurately down to extremely low power levels. Meters featuring ST's new STPM32, STPM33 and STPM34 ICs will help utilities minimise revenue losses

and ensure consistent billing for even the most frugal customers. Although existing meters are highly accurate at current levels, typically down to 50mA, errors at lower currents can bring about, in today's low-standby era, up to megawatt-hours of lost billings across a large customer base. The STPM3 devices prevent such losses by keeping the meter accuracy down to just a few milliamps, comparable to the current drawn by an LED television in standby.

These new chips also enable the development of more economical meters that help utilities reduce their operating costs. By performing power-quality calculations on-chip, including RMS voltage and current measurement, apparent-energy computation and under-voltage/over-voltage detection, the chips can offload the meter's host processor, thus simplifying software.

STMicroelectronics, Noida
Phone: (0120) 2352999
Website: www.st.com

TOOLS & EQUIPMENT

3D printer

LBD Makers has launched two models of 3D printers, MY3D BOT and XTRON, which it claims to be fast, accurate and easy to use.

Both the models feature fused-deposition modeling technology, layer resolution up to 0.1mm, 0.4mm dia nozzle with 1.75mm dia input and multi-material (ABS/PLA) printing. MY3D BOT gives 100-micron accuracy and average speed of 55mm/second and 140mm/second travel. XTRON gives 50-micron accuracy and average speed of 65mm/second and 160mm/second travel. XTRON also has an LCD screen.

LBS Makers, Chandigarh
Phone: 9999214294, 8699175703
E-mail: contactus@makerspaceindia.com

Solar micro inverter development kit

To ease designing of rapidly-growing solar-power applications, Texas Instruments (TI) announces its C2000 solar micro inverter development kit. The kit implements a complete grid-tied solar micro inverter based around TI's C2000 Piccolo TMS320F28035 microcontroller (MCU).

Rather than linking all solar panels in an installation together to a central inverter, solar micro inverter systems place smaller 'micro' inverters at the output of each individual solar panel. This lends to many benefits, including elimination of partial shading conditions, increased system efficiency, improved reliability and greater modularity. Easing solar micro inverter design.

However, solar micro inverter also poses complex challenges to designers. Designers must grapple with non-linear characteristics of the solar panel power output, devise technical maximum power point tracking (MPPT) algorithms to maximize the energy delivery, and understand complex power stage design and control. The kit, priced at US$ 850, introduces designers to a fully suitable MCU and methodically addresses the application challenges by breaking down the development process into manageable pieces.

Texas Instruments, Bengaluru
Phone: (080) 25345455, 41381665
Website: www.ti.com/in

Pick-and-place system

Essemtec has upgraded its Pantera X-plus pick-and-place system for small batches, incorporating new features for

highly-flexible small batch assembly. The enhanced X-vision system has improved image processing for small chip components and complements the existing on-the-fly laser centring feature. The combination of fast laser centring and high-precision image processing now allows customers to achieve the best possible placement results in this machine segment.

By using a new digital image top camera, image recording of fiducials can be optimised, thereby improving the detection of the PCB alignment. The Pantera X-plus is no longer compatible with Microsoft's Windows XP operating system, but works with the latest Windows 8 version.

Essemtec, Bengaluru
Phone: 9880795227
E-mail: sales@essemtec-india.com

TEST & MEASUREMENT

Portable oscilloscopes

Agilent Technologies Inc. has introduced new high-performance portable oscilloscope series deploying next-generation oscilloscope technology. The Infiniium S-Series has the world's fastest 10-bit ADC. Compared with traditional scopes with 8-bit ADCs, the S-Series has four times the vertical resolution for precise viewing of signal detail.

Coupled with a new low-noise front end, the S-Series delivers an ENOB (effective number of bits) for the system of more than eight, the highest in the industry. The series includes bandwidths from 500MHz to 8GHz with four-channel DSO models and 16-digital-channel MSO models. Standard memory is 50Mpts per channel or 100Mpts per channel when interleaved, the deepest in the industry.

The S-Series features application-specific measurement software, fuelled by a powerful motherboard with 8GB RAM. The 38.1cm (15-inch) multi-touch capacitive display is the largest in the industry, making it easier to see results and easier to use.

Agilent Technologies, New Delhi
Phone: (011) 46237100
E-mail: tm_india@agilent.com

LEDs & LED LIGHTING

Mid-LED package

Everlight Electronics has introduced the ultra-thin 940nm top-view infrared LED

IR92-01C/L491/2R in a miniature mid-LED package. The new 940nm IR LED comes in a small 2mm×1.4mm package with an ultra-low thickness of 0.7mm. This makes it ideal for all kinds of space-constrained end products like tablets and smartphones.

Along with a low forward voltage of 1.3V at 20mA, a high output power of 25mW/sr at 70mA, and a narrow viewing angle of 45° for even higher output intensity without lens, the IR92-01C series is perfectly suited for IR data transmission applications.

Everlight Electronics, Bengaluru
Phone: (080) 41623111
E-mail: salesindia@everlight.com

Chip-on-board arrays

Philips Lumileds has launched chip-on-board (CoB) arrays for parabolic aluminised reflectors (PAR) 38 equiv-alent lamps, which achieve 10 per cent or greater efficacy than competing solutions, according to the company. Also ideal for spotlights, the Luxeon CoB 1202 has a typical efficacy of 115lm/W, which varies from 95-130lm/W depending on the colour temperature and CRI of the array. It is available in a single 3-step as well as a single 5-step MacAdam Ellipse, ensuring uniform optical performance in the application.

Ideal applications include downlights and directional lamps. The arrays are all hot-tested at 85°C, which means that the luminaire design is simplified and testing can be minimised.

Philips Lumileds India distributor: Key Operation and Electrocomponents Pvt Ltd, New Delhi
Phone: 8377806486
Website: www.philipslumileds.com

POWER SUPPLIES

Tubular batteries

Base Batteries has launched Base Tuff-BT 500 tall tubular batteries for home UPS application. These are the energy-efficient backup units designed to protect homes from experiencing the inconvenience caused by power outages, minimise energy consumption and help the consumer save on electricity bills. The tall container allows for greater amount of electrolyte, which reduces maintenance and increases life even in tough power conditions.

BT series is environment-friendly and free from acid fumes, indicates water level and has a special tubular positive-plate design for long life in deep-discharge cycle.

Base Batteries, Bengaluru
Phone: (080) 41635909, 40621900
E-mail: helpline@basecorporation.com

Inspired by EFY Article!

We got some good response from the 'Automate Your Home, With Just a Switch!' article published in April issue under Innovation section. The most interesting response was from an IIMB graduate (one of the youngest) who has started a company for smart home solutions. He said, our product was the one he was looking for, and he completely agrees with the information in the article. He has started working with us as a distributor.

Syam Madanapalli
iRam Technologies
Through e-mail

MATLAB projects

I am a great fan of EFY. I used to make small projects for B.E. and M.Tech students, and now I am myself in M.Tech final year. I am grateful to EFY. Please send me a list of MATLAB-based projects.

Pavan Kumar
Bengaluru

EFY: You may refer to the following projects published in recent issues:

1. Resistor Value Calculation Standalone Application With MATLAB (April 2014)

ERRATA

In my 'Signs of Coming of Next-Gen FPGAs' article published in May issue, the second last line: "Qualcomm, which has developed such a processor together with IBM, plans to release it commercially sometime this year" is wrong. It should be read as "Qualcomm, which has also developed such a processor, plans to release it commercially sometime this year."

— Janani Gopalakrishnan Vikram,
The author

❏ In 'Resistor Value Calculation Standalone Application with MATLAB' article published in April issue, the path for the images (bound2.jpg and gray.jpg) needs to be changed as per their location in your computer.

Anjum Meer
Through e-mail

'Spot An Error' Award Winners

In eStyle First Look section in May issue, under 'Asus Launches Fonepad7 Dual-Sim Tablet,' the specifications mentioned are of Asus Fonepad 7 single-sim tablet (ME372CG) instead of Asus Fonepad 7 dual-sim tablet (ME175CG). The correct specifications are Android 4.3, 1.2GHz dual-core processor and battery capacity of 3910mAh.

Siddharth Kaul
Through e-mail

❏ In eStyle First Look section in May issue, under 'HTC One M8 Launched in India,' it is wrongly mentioned that the phone runs on a 2.3GHz Snapdragon 801 processor. Instead, it runs on 2.5GHz Snapdragon 801 processor.

Siddharth Kaul
Through e-mail

2. Colour-Sensing Robot With MATLAB (February 2014)

3. Digital Communication Design Using MATLAB (December 2013)

4. Edge Detection During Image Processing Using MATLAB (November 2013)

5. 555 Timer Design Using MATLAB (July 2013)

6. Graphics Object Counter Simulation Using MATLAB (January 2013)

To get any of these issues, contact Kits'n'Spares (email: info@kitsnspares.com)

EFY DVD

I am an avid reader of your magazine. TinyCad is a program for drawing electrical circuit diagrams, supports standard and custom symbol libraries. It supports PCB layout programs with several netlist formats and can also produce SPICE simulation netlists. It is also often used to draw one-line diagrams, block diagrams and presentation drawings. If you find it useful please include it in EFY DVD.

Prosenjeet Sengupta
Through e-mail

EFY: Thanks for your suggestion! Such feedback is very helpful. However, we

From electronicsforu.com: Circuits

The 'Voice Recorder and Playback System' construction project published in October 2013 issue is a good project. Which PCB design software was used for this circuit?

Rohit

EFY: We use gEDA software for designing PCB layouts and circuits.

❏ Is this a one-time voice recorder chip? Can we erase the voice or overwrite with new one?

Ajinkya

EFY: It is not a one-time voice recorder. You can overwrite the previous recorded message.

❏ How to play the recorded message continuously in loop?

Mohan

EFY: If you connect switch S2 (Play) permanently to ground, the recorded message will play repeatedly.

have already featured TinyCad in the DVD accompanying January issue of EFY Plus. Your further feedback would be awaited eagerly nevertheless.

Creating e-Books

The process described in 'Create e-Books from Wiki Pages' DIY article in December 2013 issue is unsuitable for those having slow Internet connection. I tried this and the e-Book created in pdf format was around 7MB. I could not download the file due to my slow Internet connection and also because the Wikipedia server is always very busy. So suggest some better ideas.

Amit Ghosh
Kolkata

The author Hari Om Prakash replies:
As far as slow Internet connection is concerned, like Internet provided by GPRS (10kbps) technology on 2G network, Wikipedia provides another option. As mentioned in the article, you can create Wikipedia user account and save the e-Book with the account so that, in future, when you are connected with a high-speed Internet connection, you can download your e-Books.

Q&A

Q1. How do we decide on the bandwidth of an oscilloscope? My company wants to buy some oscilloscopes for a new project.

Gaurang Kaparwan
Through e-mail

A1. Oscilloscope's bandwidth determines the maximum frequency signal that it can measure accurately. The accuracy decreases with increase in signal's frequency. The bandwidth mentioned in the datasheet (say, 100MHz) is actually the frequency at which a sinusoidal input is attenuated to 70.7 per cent of its true amplitude. Beyond this frequency the oscilloscope cannot support reasonable accuracy.

To decide how much bandwidth you need, find out the range of frequencies you would need to measure. Once you know the range, just use the Five Times rule. Multiply the maximum frequency by five and you have the bandwidth that you would need to accurately measure those signals. For example, to measure signals up to 10MHz frequency accurately, you would need an oscilloscope with 50MHz bandwidth. Otherwise, the high frequency changes will not be resolved and the amplitude will be distorted.

In case of digital signals, it is important to capture the fundamental, third and fifth harmonics to display accurate results. Therefore the bandwidth of the scope, together with the probe, should be at least five times the frequency of the measured signal.

Q2. I have been learning electronics by trying your circuits at home. The circuit diagrams are so neat and pleasing to eyes. What circuit design software do you use?

Imran Ali
Through e-mail

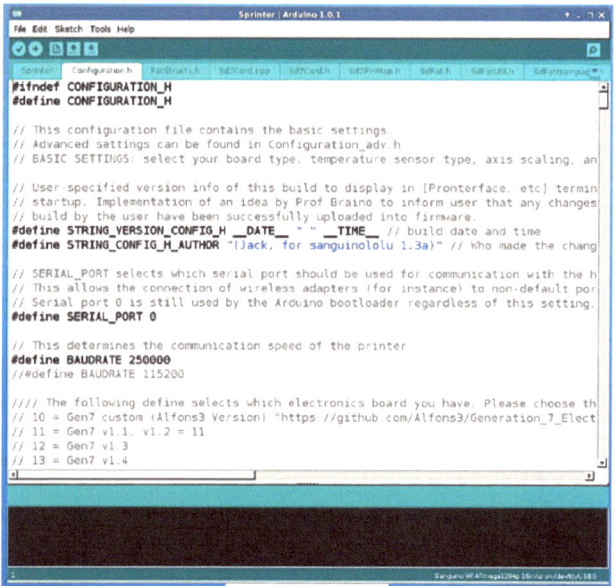

Configuration.h file in Arduino IDE

A2. We use gEDA Project open source software to make the circuit diagrams and PCB drawings that are published in the magazine. The software has all the tools for complete product design. As it is open source, you do not have to pay for license. Also, there is no limitation like number of layers in PCB design.

However, you will not be able to create the exact schematic that we publish in the magazine. We have customised the software to meet our requirements of background colour, line colours, line width, etc. You can also do the customisation to suit your taste or requirement. But you can also use the software without any customisation, if you wish.

Help is available from those who are already using gEDA. To find out all the resources, check:

http://www.geda-project.org/

Q3. I am trying to work out a 3D printer with Sanguino board. The boot loader is pre-installed but I am unable to flash the software (sprinter.pde). Please help with the complete steps.

Dinesh Rai
Through e-mail

A3. Your board came pre-installed with the boot loader, but I would suggest that you reinstall it using the tutorial at http://reprap.org/wiki/Burning_the_Sanguino_Bootloader_using_Arduino_as_ISP to be double sure. Once the boot loader is installed, follow below-mentioned steps to flash sprinter.pde in the board using Ubuntu operating system:

1. Download Sprinter software from https://github.com/kliment/Sprinter.git

2. Arduino IDE will be needed to compile and program the board. Install required packages using the command:

```
$ sudo apt-get install arduino-core
```

3. Get Arduino software version 0023 and extract it in a directory. Get Sanguino software from http://sanguino.cc/softwareforlinux and copy it in Arduino-0023/hardware/directory.

4. Open Makefile in the downloaded Sprinter folder and edit the INSTALL_DIR. Save and close the file.

5. Open the terminal and go to the Sprinter directory. Type 'make.' If everything goes well, Sprinter/applet/Sprinter.cpp should get created.

6. Connect your Sanguinololu to your computer, launch arduino-0023/arduino and open Sprinter/Sprinter.pde. Select the correct COM port and board.

7. Click configuration.h file in the Arduino IDE and edit Motherboard and other things to suit your setup. Once done, click the upload button and the program will flash in the board.

Answers compiled by EFY technical editor, Ankit Gupta. Letters and questions for publication may be addressed to Editor, Electronics For You, D-87/1, Okhla Industrial Area, Phase 1, New Delhi 110020 (E-mail: editsec@efyindia.com) and should include name and address of the sender

KAPIL SOOD
MANAGING DIRECTOR,
TEKTRONIX INDIA

"Time to market is a critical factor for design engineers"

The new oscilloscopes are evolving in multiple areas, but the most interesting aspect seems to be the increased focus on the user— the engineer who actually uses the scope. Kapil Sood, managing director, and Naresh Narasimhan, country manager at Tektronix India spoke with Dilin Anand of EFY

NARESH NARASIMHAN
COUNTRY MANAGER,
TEKTRONIX INDIA

What are the most exciting changes affecting scopes recently?

The oscilloscopes market is characterised by three distinct categories—basic, main and professional. In the mainstream segment, users are mostly embedded engineers working on a mixed-domain design. In the analogue-to-digital era, a lot of digital designs came onto the analogue board, thus bringing up the Mixed Signal Oscilloscopes (MSOs) several years ago. The trend has evolved since, and what we are seeing today is that this mixed domain has moved to another level with the inclusion of wireless RF signals. Analogue, digital and now RF signals are coming together to the embedded domain, creating significant changes on how work is done in designs houses and research centres, specifically with respect to analysing those signals in as short a time as possible.

How is this affecting engineers' design work?

RF signals are also now being added to the mix of digital and analogue signals that have to be analysed. This is not easy for the design engineers working on the project, because they have to put all of these together and analyse on multiple fronts to successfully debug and complete a design. Time to market is a critical factor for design engineers, since the product lifecycle period is significantly shorter today. It means they have to come out with a newer version of their design three months down the line, which also results in a shorter innovation cycle. Thus the engineers need tools where time to answer is shorter for them.

What are the elements that such a tool should have?

With integrated instruments, engineers get multiple tools in one instrument in a compact form factor, and are thus able to get results faster. The key aspect of the integration in an MDO3000 is the inclusion of a logic analyser, protocol analyser, spectrum analyser, function generator and digital voltmeter into the oscilloscope itself. This is cru-cial from the customers' perspective. Their tool of choice to debug and verify their designs is the oscilloscope. The other instruments are used on an intermittent basis.

So is it relevant only to those working on debugging on multiple domains?

Multiple instruments are not only about the ability to debug in multiple domains, it's also about a design engineer who has the time to debug thoroughly. With a mixed-domain oscilloscope, he is able to use two methodologies to debug—the frequency domain approach or the time domain approach. Previously, he could use only time domain. With two options to debug or validate his design, the chances of his design passing through or his bringing the product into market, is much higher. Instruments like these are relevant to a broader group of design engineers, both education and research people. The multiuse features of MDO3000, for instance, offer the functionality of six independent instruments without the significant cost of separate instruments.

How exactly is this instrument beneficial for a design team starting out?

The basic customer will continue to use the basic scope as even design houses still have a combination of basic, mid-stream and performance scopes. When you look at the ROI aspect, there is an expectation that the upgradability built into the instrument should make it scalable for the future. In fact, the upgradability of instruments like this new mixed-domain oscilloscope sets it apart in the marketplace due to both the number of upgrades available as well as how easy the upgrade process is. There are two major categories of upgrades. First, there are performance upgrades. Customers can upgrade the bandwidth of their oscilloscope all the way up to 1GHz. When the analogue bandwidth is upgraded, the frequency range of the included spectrum analyser is also upgraded to the new bandwidth level.

There are also functionality upgrades. These include adding the digital channels, arbitrary function generator, a variety of serial trigger/decode solutions as well as other advanced analysis packages, such as power analysis and limit/mask testing.

What are the different challenges that are solved by such an instrument?

Sometimes, when you debug, you do not find your point of interest or fault on your device in the time domain. The issue would be in the frequency domain, but because it is not visible in time domain the fault lies hidden. In such cases, it should be possible to change to frequency domain easily. And that's when design engineers pull over the spectrum analyser. Later, when the software and hardware come together, you understand what's happening to the protocol and have to co-relate every code to a physical signal, which is another spectrum analyser or the logic analyser of the scope. Every design house is doing something on this level of design, as this is how it comes together in the workflow. As long as there is a need to lower the time to market for a product, the trend will be seen going towards multiple scalable instruments, and design engineers will need the ability to get work done at one shot.

What else is being done to improve users' experience with instruments?

Another important aspect is usability and the user interface. If I am comfortable in using an oscilloscope, it might not be easy for me to use a different instrument, because the user interface, the contrast and the pattern that gets formed might be different. A crucial thing here is that engineers can spend more time on their design, because they don't have to spend time learning the instrument.

Will instruments like these make sense for start-ups and small design houses running on shoestring budgets?

When a design house starts up in Bengaluru, they might have just two or three people to begin with. Later, as the company hires more people and begins to grow, they will need instruments that grow with them. So rather than buying a new expensive instrument, they can simply upgrade their current basic instrument.

For low-power designs, any functionality or features that aim to help?

There is a power module that goes into some instruments. It aims to help the users with just that. Engineers who want to improve switching power supply efficiency can use it to measure power loss at the switching device. When working on low-power designs, one has to depend on the power supply that typically moves from linear to switching in order to be power efficient. Power modules like these allow you to look at the rise time and fall time of the switching, and can be used to evaluate various options like topology. ●

"What should you look for in a chip?"

Which chip do you select for high-performance parallel calculations and control applications? How do you reduce the power consumption on a chip? Satish Bagalkotkar, president and CEO, and Devesh Gautam, COO, Synapse Design answer the above questions and more as they speak with Dilin Anand of EFY at Synapse's new office in Whitefield

SATISH BAGALKOTKAR
PRESIDENT AND CEO,
SYNAPSE DESIGN

DEVESH GAUTAM
COO,
SYNAPSE DESIGN

What are the most significant technological advances that enable reducing power consumption in a chip?

They are advanced grain power gating, fine grain clock-gating techniques, RTL modification for clock gating (EDA tool based) and power-aware architecture of the RTL (RTL structures, data path parallelisation and optimisation of the code in terms of VT usage). Overall, we want to reduce the toggle rate of heavy load nodes with RTL changes or special custom cell design. In a nutshell, design for multipower domain based on an application instead of the whole circuit all the times.

What are the latest techniques for getting the chip to work at a lower power level in the real world?

Before we start most low-power designs, we review the product specifications and physical cell libraries and memory macros. This is to reduce library cells that use deep stacks and memory macros with small timing and voltage margin.

Since you have an expertise in CPU and GPU architectures, could you give us examples of the applications where a GPU would be a better choice than a CPU?

GPUs are optimal for tight code with high parallelisation. They have hundreds of simpler execution units with single instruction, multiple data (SIMD) capabilities that enable algorithms that can be parallelised easily to be best implemented on GPUs. At the same time, the GPU normally has smaller memory availability in normal implementation than a CPU, so a GPU is probably not well suited for a data-intensive application.

What makes the GPU better at the applications mentioned above than a CPU?

Applications listed above produce a lot of data that have to be processed very fast. GPU is mainly used as an accelerator for SIMD processing and CPU is used as a controller. Architecturally, most of the GPU blocks (shader and rendering engine) are used in parallel for faster operation.

What is the primary difference in working between a CPU and a GPU—with respect to the analogy that a CPU is like being an executive while a GPU is a labourer?

Most CPUs have an execution unit with branch and loops capability that is used for controlling other logic blocks. As an

analogy, we can equate CPU as an executive or boss and cores as labourers. GPUs are meant for highly parallel calculation, where data streams are operated upon, not for control application like a normal CPU is.

What should you look for in a chip?

Some features that apply to many industries are low power and performance, design of testability and manufacturability, hardened blocks for best speed, area and power (PPA)—ARM processor, GPU shader, network processor core, security engine and interface blocks—PCIe and Serdes and mixed signal design.

Since you are also into FPGA, does the traditional FPGA value proposition of reduced time-to-market, lower development cost (total) and flexibility completely apply in the test and measurement world too?

I personally think it does, because it allows the smaller production runs that some instrumentation companies need. However, speed and capacity are still a problem. Very high performance scopes, for example, have a data throughput need that is simply not compatible with FPGAs. I once worked on an ASIC for a scope that needed 1.5Gbit of embedded memory, with a total throughput of 200Gbps in and out of the chip or one that requires complex filters and FFT processing.

What approach is taken by SoC designers to tackle the greater share of analogue/mixed-signal circuitry in the chip—especially since analogue does not scale as well as digital.

At the block level, we enforce a methodology using thorough transistor level simulation at all the PVT corners. We use silicon-verified sub-blocks as much as possible, implement the design for testability for analogue blocks and verify all the block-level interfaces between analogue and digital circuits. A thorough design review is held with the customer and block owners.

In verification, we help our customers with writing the verification models and implementing an efficient verification methodology, such as AMS Verilog. ●

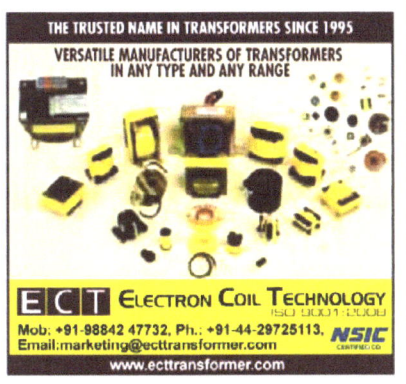

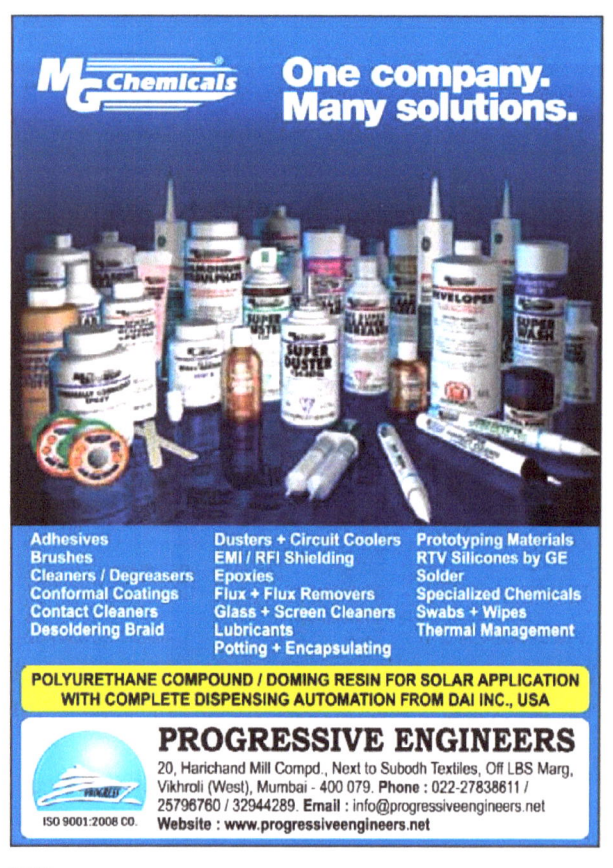
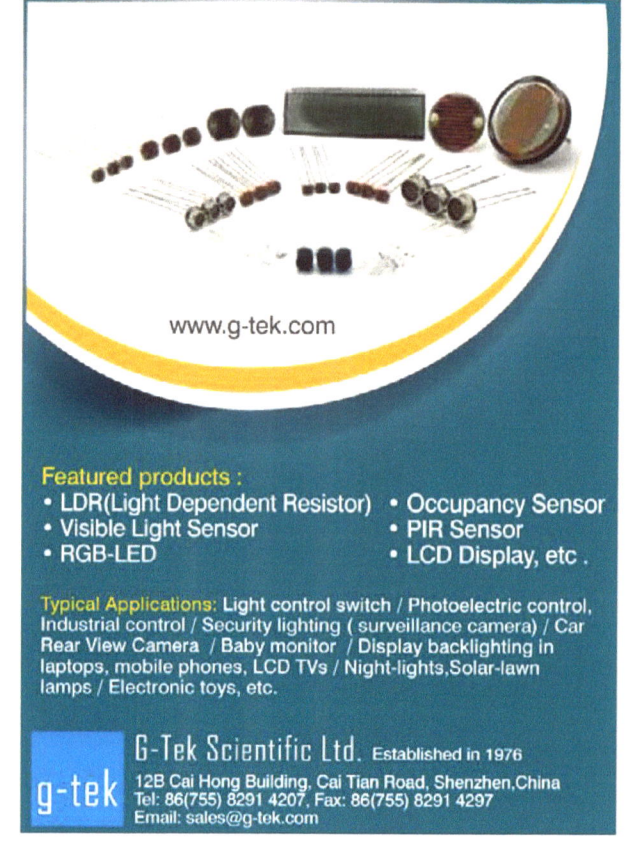

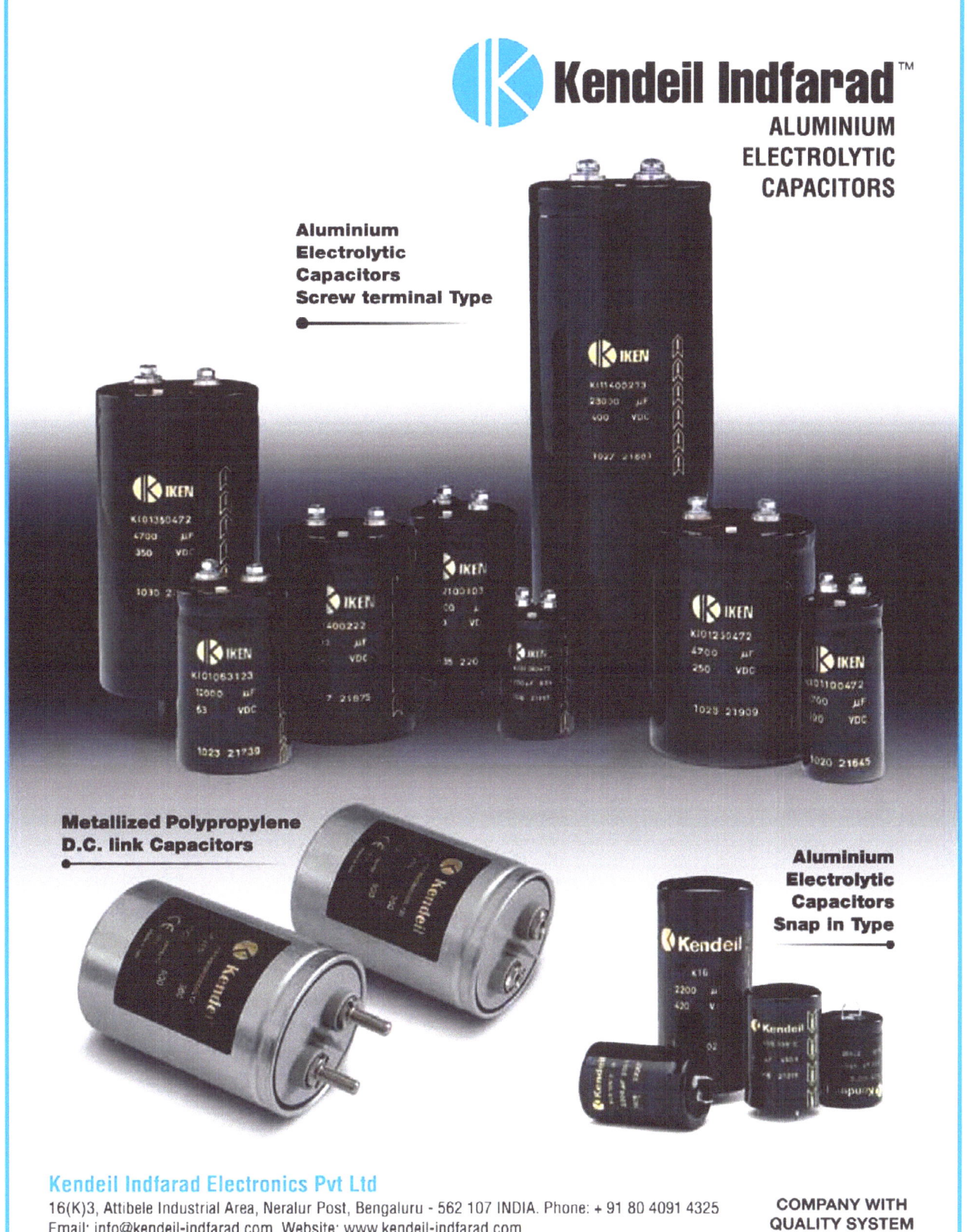

Kendeil Indfarad™
ALUMINIUM ELECTROLYTIC CAPACITORS

Aluminium Electrolytic Capacitors Screw terminal Type

Metallized Polypropylene D.C. link Capacitors

Aluminium Electrolytic Capacitors Snap in Type

Kendeil Indfarad Electronics Pvt Ltd

16(K)3, Attibele Industrial Area, Neralur Post, Bengaluru - 562 107 INDIA. Phone: + 91 80 4091 4325
Email: info@kendeil-indfarad.com, Website: www.kendeil-indfarad.com

A MEHER + Kendeil Joint Venture

COMPANY WITH
QUALITY SYSTEM
CERTIFIED BY DNV
= ISO 9001:2008 =

electronicsrocks.com

An **EFY**GROUP EVENT

call for
papers

EFY invites experts to speak at Electronics Rocks Conference in Bengaluru on October 10 and 11, 2014.

Guidelines

1. One speaker may submit up to 3 applications

2. Speakers should be professional engineers, though EFY does not mandate a minimum number of years of experience

3. Submissions should be made before June 15, 2014.

4. Submissions should be made at electronicsrocks.com/call-for-speakers

Sectors on which topics for technical sessions will be based:

- Automotive
- Medical
- Defence & Aerospace
- Rapid Prototyping and 3D Printing
- Internet of Things

Submit applications at
electronicsrocks.com/call-for-speakers

3rd Edition

electronics
Rocks 2014

October 10-11, NIMHANS, Bengaluru

Workshop Partners  COMSOL  Atmel | Session Partner  GOPEL electronics

www.efytechcenter.com

Learn. Lead. Succeed.

Train with EFY Tech Centers

- DELHI • HYDERABAD • KOLKATA
- GWALIOR • COIMBATORE* NEW

(* For inhouse Trainings in educational institutes)

REGISTRATIONS OPEN FOR SUMMER TRAINING

ENJOY 10% DISCOUNT... by booking in advance with payment, at least 15 days before a course begins.

Hands-on **Training programs of EFY For Electronics Engineers, Students & Hobbyist To Give Them An Edge Over Others In Their Field ...Make Your Move For Better/ Self Employment TODAY!**

FRANCHISEE OPPORTUNITY AVAILABLE FOR WEST & SOUTH INDIA

Contact: efytech@efyindia.com or call Deepika on 9650547770

Hands on Training	Course Descriptions
PCB Design	**Duration:** 3 Days (18 hours) **Course Fee:** Rs 3.500/- (including service tax)
Basics of Electronics	**Duration:** 3 days (18 hours) **Course Fee:** Rs 3.500/- (including service tax)
Robotics	**Duration:** 9 days (54 hours) **Course Fee:** Rs 10,500/- (including service tax)
Microchip PIC16F/ PIC18F Microcontroller	**Duration:** 6 days (36 hours) **Course Fee:** Rs 7,000/- (including service tax) • Certificate awarded by Microchip Inc., USA
Free RTOS Using PIC18F4520	**Duration:** 9 days (54 hours) **Course Fee:** Rs 10.500/- (including service tax) Knowledge of C-Programming and PIC18F Microcontroller necessary
VLSI	**Duration:** 9 days (54 hours) **Course Fee:** Rs 10.500/- (including service tax) Knowledge of Digital Circuit is necessary
Embedded Systems	**Duration:** 25 days (150 hours) **Course Fee:** Rs 31,500/- (including service tax)
Raspberry Pi NEW	**Duration:** 5 days (30 hours) **Course Fee:** Rs 10,000/- (including service tax)
Arduino NEW	**Duration:** 5 days (30 hours) **Course Fee:** Rs 10,000/- (including service tax)

The courses are concise with focus on hands-on.
Please check our website for course schedule and other details.

Trainings also conducted in companies and educational institutes

Mode of payment: The fee may be paid online (www.efytechcenter.com/course), by money order, through a DD or cheque in the name of EFY Enterprises Pvt Ltd, payable at Delhi. All payments, preferably by the above modes, to be sent *only* to H.O at Delhi
• Payment may also be made in cash at the respective center itself, either before or at the time of joining the course
• Please re-confirm the dates a week before attending. * *Refund in case of cancellation will be 50% of the fee paid.*
Accommodation can be arranged in a private hostel/ guest house on prior request, at an extra cost.

Send your queries to: efytech@efyindia.com

EFY Enterprises Pvt Ltd
D-88/3, Okhla Industrial Area, Phase 1, New Delhi 110020, Phone: 011-40596626, 09650013762, 09650547770

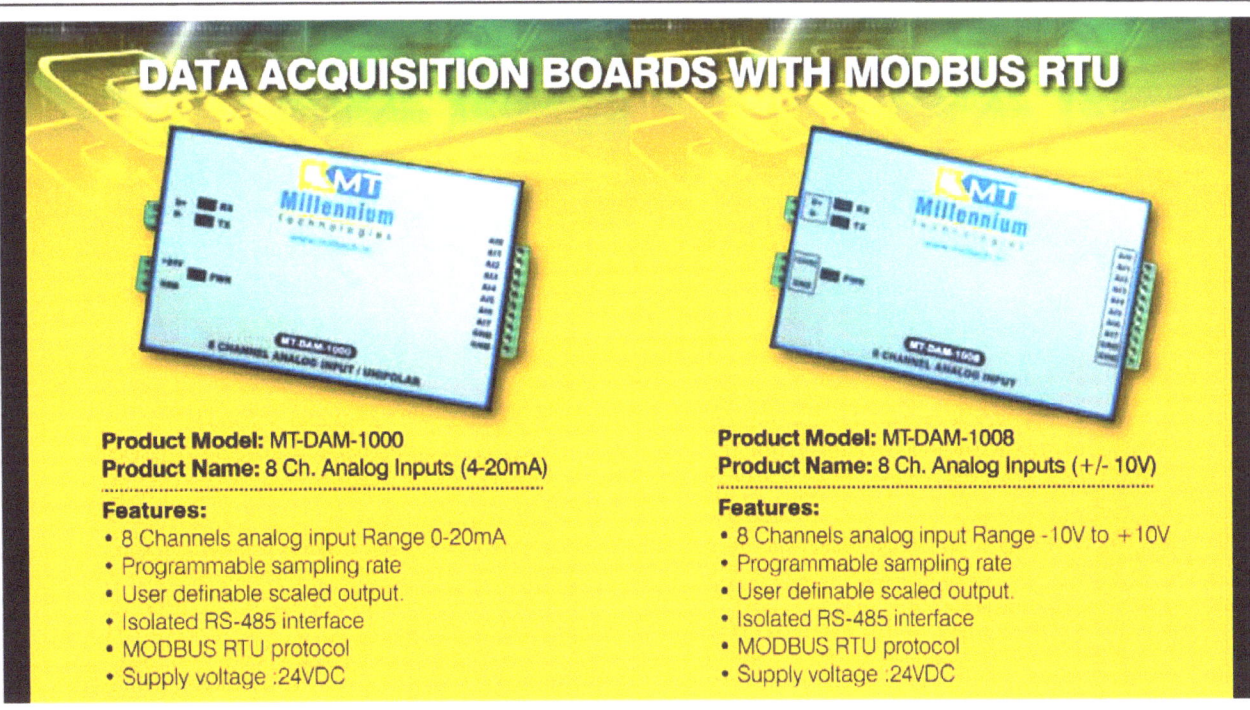

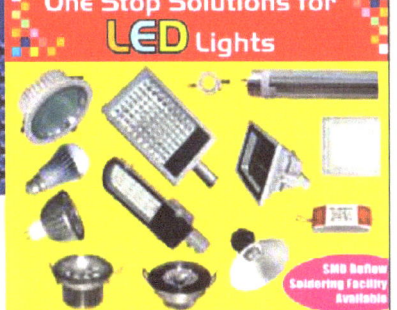
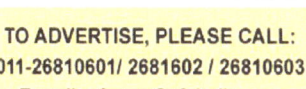

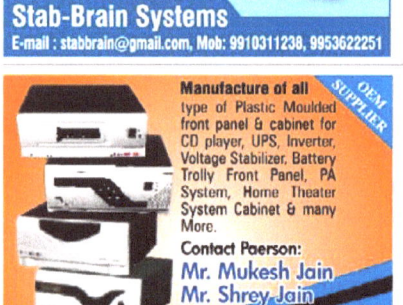

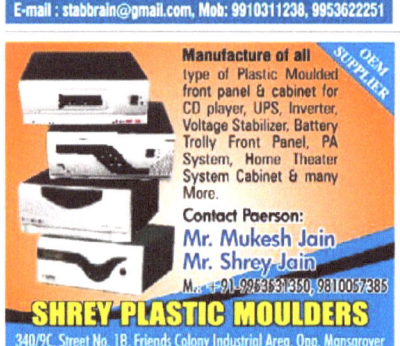

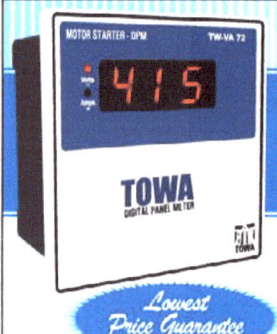

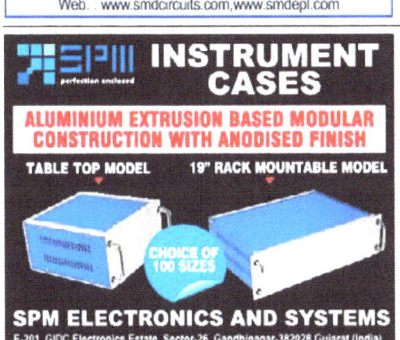
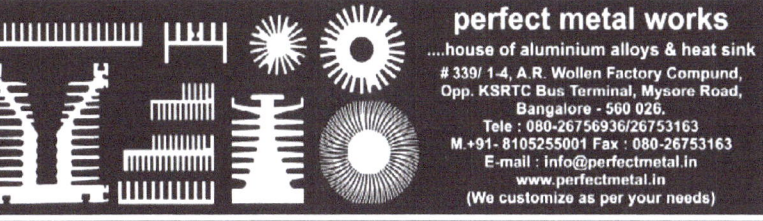

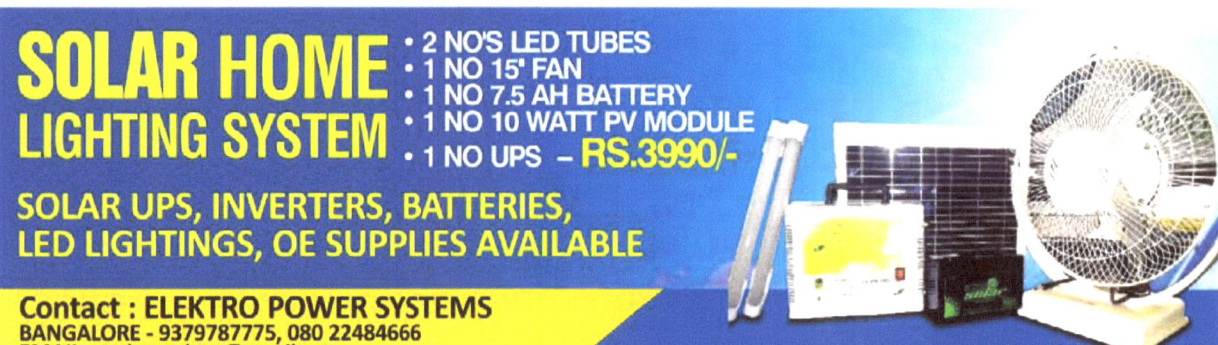
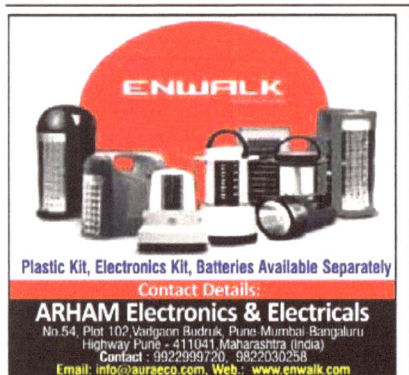

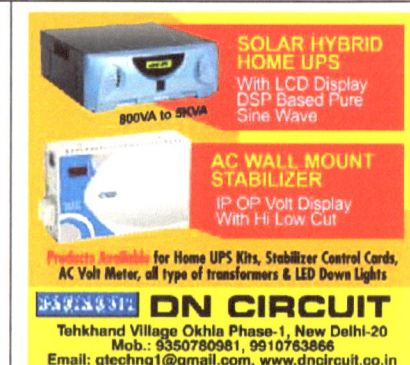

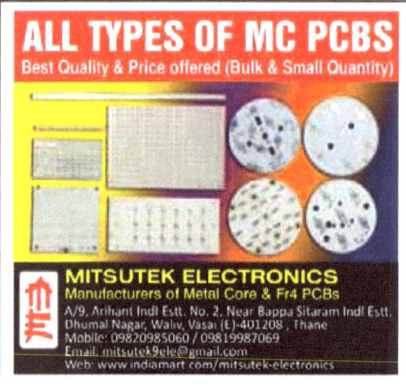

An **EFY**GROUP INITIATIVE

Access OLD ISSUES For Free!

If you are an active subscriber of any EFY publication, you can access its current and old issues for free.

REGISTER* ON HTTP://EZINE.EFY.IN TO ACCESS OLD ISSUES.

Access is available through our e-zine edition for all our publications. After your subscription expires, you will only be able to access those issues that you had subscribed to.

Access to old issues will soon be available through Android App too.
*For those who've already registered, you can access this feature through the same account.

FOR ANY DOUBTS OR QUERIES CALL +91-11-40596600 (EXTENSIONS: 201 & 202)

ADVERTISERS' PRODUCT CATEGORY INDEX

EFY Magazine Attractions During 2014

MONTH	TECHNOLOGY FOCUS	EFY REPORT	BUYERS' GUIDE	T&M
January	Electronics of Things	Industrial Automation Electronics	Rework Stations	Automated Test Equipment (AOI, etc)
February	Smart Grid	Smart Grid Electronics	Handheld T&M Equipment for Field Engineers	Thermal Imaging
March	Smart & Electric Vehicles	Automotive Electronics	How to Make Your Lab Static Proof	Function & Signal Generators
April	Smart Homes	Inverters & UPS–SOHO & Industrial	Digital Multimeters	Multimeters
May	FPGA (Programmable Chips)	Connectors & Terminals	FPGA Training Kits	Data Acquisition Systems
June	3-D Printers	Certification & Quality Labs	Desktop Manufacturing Equipment (SMT, Reflow Ovens, 3D Printers)	EMC Test Equipment
July	Raspberry Pi	PCB Industry in India: Suppliers & Manufacturers	Budget Friendly Oscilloscopes	Oscilloscopes
August	Security 2.0: Latest products	Aerospace & Defence Electronics	Wi-fi & RF Modules	Incircuit Test Systems
September	Smart Robos	Solar Electronics	EDA Tools for Circuit Design	Virtual Instruments
October	Open Source Electronics	Educational & Training Products	Development Boards (Microcontroller based)	Analysers (Network, Protocol, Spectrum, etc)
November	Wireless Communication Technologies (Zigbee, RF to 5G & beyond)	Security & Surveillance	Soldering / Desoldering Stations	RF Devices (Wireless Devices)
December	Smart Lighting	LED Lighting	Programmable Power Source	Power Analysers/Power Meters/Supplies

ADVERTISERS' INDEX

Page numbers subject to final dummy corrections

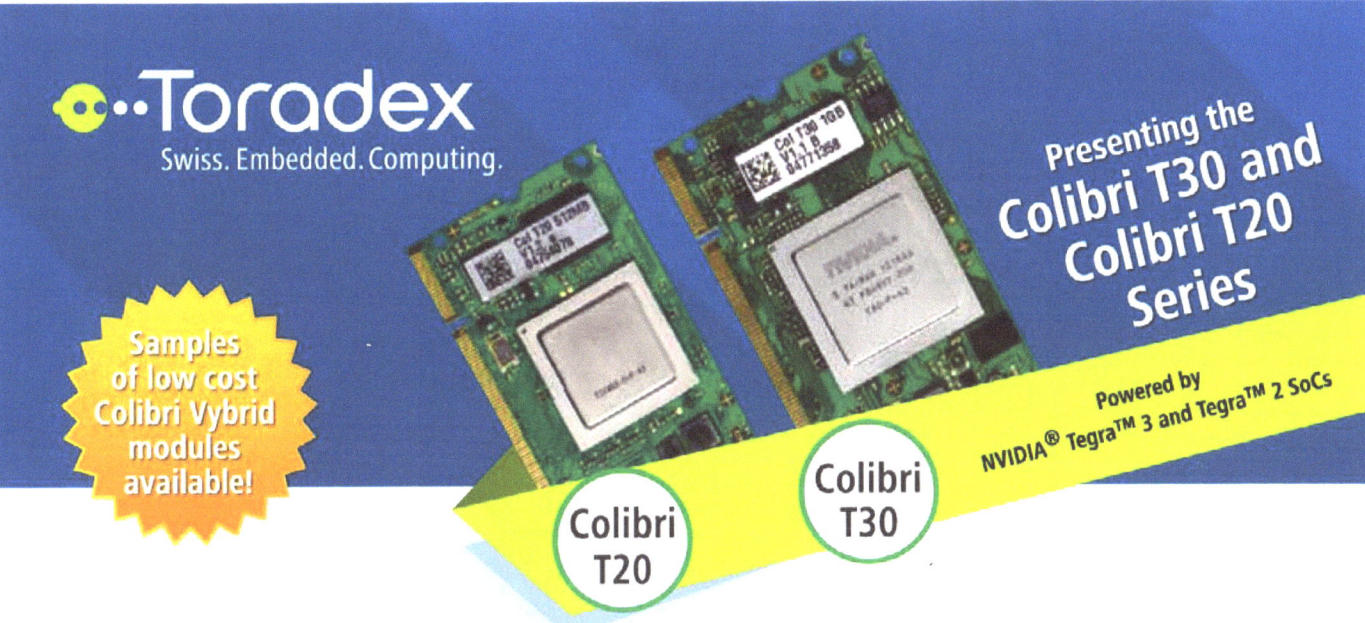

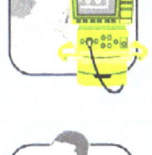

RNI No. 17587/69, Mailed on 1st/2nd of same month   Delhi Postal Regd. No.DL(S)-01/3097/2012-14
Published on 29th of advance month   Licenced to Post without Pre-Payment Licence No.U(SE)-08/2012-14

6 Instruments.1 Scope.
Infinite Versatility.

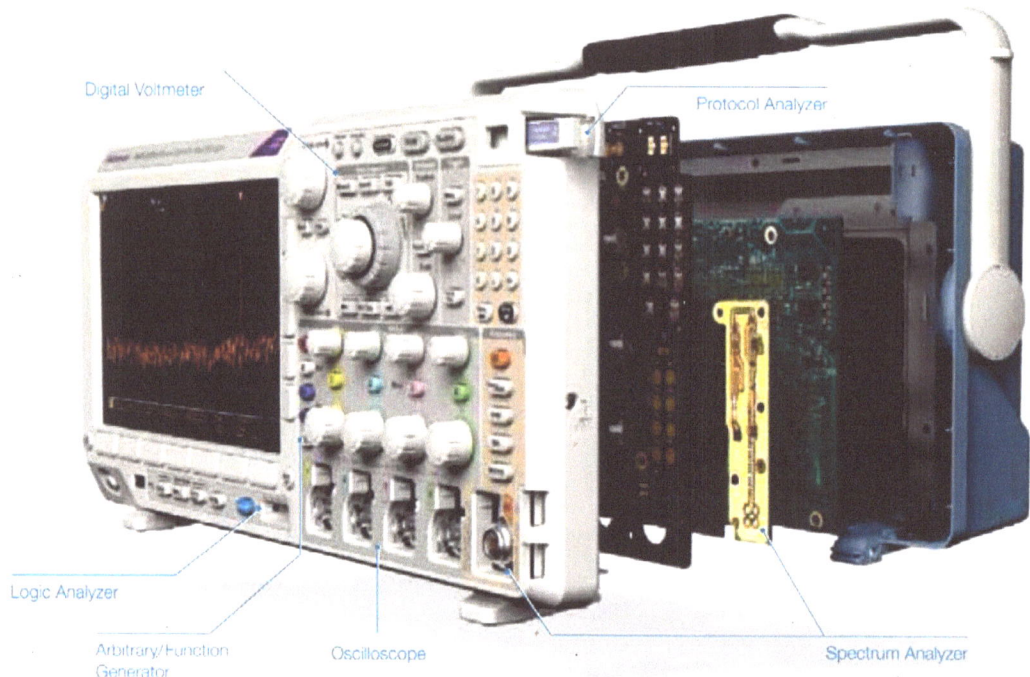

The 6-in-1 MDO3000 Mixed Domain Oscilloscope with a spectrum analyzer. Starting at $3,350.*

Designs have evolved, integrating analog, digital and RF technologies. Now, it's the oscilloscope's turn. Introducing the MDO3000 Mixed Domain Oscilloscope. Featuring a spectrum analyzer, function generator and more, it's the only scope that comes with the instruments you use most, built-in. It's also upgradeable, enabling you to add instruments and increase bandwidth as your needs grow. So you get unlimited versatility at a starting price that's anything but limiting.

6 Advanced Instruments Built In

- Oscilloscope: 100 MHz-1 GHz models, 2 or 4 channels, 5 GS/s sample rate, 10 Mpoint record length
- Spectrum Analyzer: 9 kHz-3 GHz with 3 GHz wide capture bandwidth
- Logic Analyzer: 16 channels, 121 ps timing resolution
- Arbitrary/Function Generator: 50 MHz, 13 standard plus arbitrary waveforms
- Protocol Analyzer: Serial data trigger, decode and analysis
- Digital Voltmeter: 4-digits resolution

For more information, please call our sales channel partner in the city nearest to you. **Ahmedabad :** Optimized Solutions Pvt. Ltd. (91-79) 30080808 / 26589008; **Bangalore:** Aarjay International (91-80) 43409200, Convergent Technologies (91-80) 23490111 / 179; **Chennai :** Aarjay (91-44) 40509200, Primetech Instruments Pvt. Ltd. (91-44) 24492961; **Andhra Pradesh / Orissa :** Peridot Technologies (91-40) 23405855; **Indore :** Optimized Solutions Pvt. Ltd. (91) 9909979950; **Kerala :** Convergent Technologies (91-471) 2462412 / 386; **Kolkata :** Techno Scientific : (91-33) 32948567; **Mumbai :** Vitronics India (91-22) 28506037; **New Delhi :** S.P.I. Engineers Pvt. Ltd. (91) 9810157421; Convergent Technologies (91-11) 42481121 / 1131; **Pune :** Cyronics Instruments Pvt. Ltd. (91-20) 24208200; **North East India :** Vishal Vyapar Vikash (91-361) 2540701.

Enquires Email To : india.mktg@tek.com